Jing Zhou

# Feature Selection
# in Data Mining

I0009602

Jing Zhou

# Feature Selection in Data Mining

## Approaches Based on Information Theory

VDM Verlag Dr. Müller

Bibliographic information by the German National Library: The German National Library lists this publication at the German National Bibliography; detailed bibliographic information is available on the Internet at http://dnb.d-nb.de.

This works including all its parts is protected by copyright. Any utilization falling outside the narrow scope of the German Copyright Act without prior consent of the publishing house is prohibited and may be subject to prosecution. This applies especially to duplication, translations, microfilming and storage and processing in electronic systems.

Any brand names and product names mentioned in this book are subject to trademark, brand or patent protection and are trademarks or registered trademarks of their respective holders. The use of brand names, product names, common names, trade names, product descriptions etc. even without a particular marking in this works is in no way to be construed to mean that such names may be regarded as unrestricted in respect of trademark and brand protection legislation and could thus be used by anyone.

Copyright © 2007 VDM Verlag Dr. Müller e. K. and licensors
All rights reserved. Saarbrücken 2007
Contact: info@vdm-verlag.de
Cover image: www.purestockx.com
Publisher: VDM Verlag Dr. Müller e. K., Dudweiler Landstr. 125 a, 66123 Saarbrücken, Germany
Produced by: Lightning Source Inc., La Vergne, Tennessee/USA
　　　　　　Lightning Source UK Ltd., Milton Keynes, UK

Bibliografische Information der Deutschen Nationalbibliothek: Die Deutsche Nationalbibliothek verzeichnet diese Publikation in der Deutschen Nationalbibliografie; detaillierte bibliografische Daten sind im Internet über http://dnb.d-nb.de abrufbar.

Das Werk ist einschließlich aller seiner Teile urheberrechtlich geschützt. Jede Verwertung außerhalb der engen Grenzen des Urheberrechtsgesetzes ist ohne Zustimmung des Verlages unzulässig und strafbar. Das gilt insbesondere für Vervielfältigungen, Übersetzungen, Mikroverfilmungen und die Einspeicherung und Verarbeitung in elektronischen Systemen.

Alle in diesem Buch genannten Marken und Produktnamen unterliegen warenzeichen-, marken- oder patentrechtlichem Schutz bzw. sind Warenzeichen oder eingetragene Warenzeichen der jeweiligen Inhaber. Die Wiedergabe von Marken, Produktnamen, Gebrauchsnamen, Handelsnamen, Warenbezeichnungen u.s.w. in diesem Werk berechtigt auch ohne besondere Kennzeichnung nicht zu der Annahme, dass solche Namen im Sinne der Warenzeichen- und Markenschutzgesetzgebung als frei zu betrachten wären und daher von jedermann benutzt werden dürften.

Copyright © 2007 VDM Verlag Dr. Müller e. K. und Lizenzgeber
Alle Rechte vorbehalten. Saarbrücken 2007
Kontakt: info@vdm-verlag.de
Coverbild: www.purestockx.com
Verlag: VDM Verlag Dr. Müller e. K., Dudweiler Landstr. 125 a, 66123 Saarbrücken, Deutschland
Herstellung: Lightning Source Inc., La Vergne, Tennessee/USA
　　　　　　Lightning Source UK Ltd., Milton Keynes, UK

ISBN: 978-3-8364-2711-1

ABSTRACT

STREAMWISE FEATURE SELECTION

Jing Zhou

In streamwise feature selection, new features are sequentially considered for addition to a predictive model. When the space of potential features is large, streamwise
feature selection offers many advantages over traditional feature selection methods,
which assume that all features are known in advance. Features can be generated
dynamically, focusing the search for new features on promising subspaces, and overfitting can be controlled by dynamically adjusting the threshold for adding features
to the model. In contrast to traditional forward feature selection algorithms such as
stepwise regression in which at each step all possible features are evaluated and the
best one is selected, streamwise feature selection only evaluates each feature once
when it is generated. We describe information-investing and $\alpha$-investing, two adaptive complexity penalty methods for streamwise feature selection which dynamically
adjust the threshold on the error reduction required for adding a new feature. These
two methods give false discovery rate style guarantees against overfitting. They differ
from standard penalty methods such as AIC, BIC and RIC, which always drastically
over- or under-fit in the limit of infinite numbers of non-predictive features. Empirical results show that streamwise regression is competitive with (on small data sets)
and superior to (on large data sets) much more compute-intensive feature selection
methods such as stepwise regression, and allows feature selection on problems with
millions of potential features.

When doing feature selection in multiple simultaneous regressions, one can "borrow strength" across the different regressions to get a more sensitive criterion for
deciding which features to include in which regressions. We use information theory
to derive the Multiple Inclusion Criterion (MIC), an efficient coding scheme, in both
stepwise and streamwise feature selection. Each feature can be added to none, some,
or all of the regression models. Experiments show that the MIC approach is useful

for selecting a small set of features when predicting multiple responses from the same set of potential features.

COPYRIGHT

Jing Zhou

2006

# Contents

# List of Tables

ix

# List of Figures

# Chapter 1

# Introduction

## 1.1 Streamwise Feature Selection

Building accurate predictive models in many domains requires consideration of hundreds of thousands or millions of features. Models which use the entire set of features will almost certainly overfit on future data sets. Standard statistical and machine learning methods such as SVMs, maximum entropy methods, decision trees and neural networks generally assume that all features are known in advance. They then use regularization (e.g. ridge regression, weight shrinkage in neural networks, or Gaussian or other priors), or features selection (e.g. tests on hold-out data, or penalty methods such as AIC, BIC, RIC, or FDR) to avoid overfitting. There is a large class of problems in which feature selection gives higher prediction accuracy than smoothing.

Feature selection offers many benefits beyond improved prediction accuracy. For example, feature selection can reduce the requirements of measuring and storing data. Non-predictive or redundant expensive features may not need to be collected or captured. Feature selection can also speed up model updating. When large amount of new information are available on a minute-by-minute basis, the time to update models can be prohibitive if large sets of features are involved. Selecting the

1

most important features from the new information will speed up the model updating. Other needs include better data understanding. The book will focus on selecting a subset of features to build a good predictive model.

If we consider only the $m$ raw features, number of possible feature subsets is $2^m$. Searching over all possible models is, of course, intractable. The problem is exacerbated when interaction terms are considered. Stepwise regression adopts the greedy search method: hill climbing. This method considers all features at every iteration, the best one is selected, and then all remaining features considered, etc. Stepwise selection is terminated when either all candidate features have been added, or none of the remaining features lead to increased expected benefit according to some measure, such as a p-value threshold. Stepwise regression requires having a finite set of features, and requires looking at each features many times. This is computational expensive, and, in some of our experiments, stepwise package in R didn't work due to memory overflow.

This book presents a method we call streamwise feature selection, which does feature selection from a stream of features and each feature is considered only once. As each feature is observed, it is tested for inclusion in the model and then either included or discarded. This approach offers many advantages over the traditional approach of considering a fixed set of features, such as stepwise regression. For linear and logistic regression, we have found that streamwise regression can easily handle millions of features. More details of streamwise feature selection are provided below.

Of the vast number of candidate features, often most are spurious; only a small number are expected to be useful in building a predictive model. Even among the small number of useful features, some features are redundant to the others. Building predictive models from such large, complex data sets requires careful control to avoid overfitting, particularly when there are many more features than observations.

Any feature selection procedure occasionally generates a false positive. That is, the procedure accepts the features as useful ones that either offers no improvement or degrades the accuracy of the model. If one has a possible infinite sequence of candidate features from which to choose, this property could produce serious over-fitting. When presented with a continous sequence of features that are random noise, a selection procedure that generates false positives at a fixed rate will select infinitely many of these random features.

A variety of different methods have been used over the years to avoid overfitting when doing feature selection. Cross-validation on a held out data set is popular in machine learning, but has the disadvantage over penalty methods such as AIC, BIC and RIC of requiring either enough data to hold out (e.g., half the data) or enough computer time to do (e.g., 10-fold) cross-validation. This book focus on penalty-based feature selection methods for problems in which a small number of predictive features are to be selected from a large set of potential features.

The assumption behind penalty methods such as AIC, BIC, and RIC are not met when a fixed number of features are to be selected from an arbitrarily large set of potentially predictive features. Inclusion rules such as AIC and BIC, which are not a function of $m$, the number of possible features to be considered for inclusion in the model, inevitably overfit as $m$ becomes large. Inclusion rules such as RIC or Bonferroni correction which are a function of $m$ underfit as $m$ becomes large. Any such method that reduces the chance of including each feature based on the total number of features, $m$, to be considered will end up not adding any features in the limit as $m \to \infty$.

The solution to this dilemma is to devise a method which incrementally adjusts the criterion for including new features in the model depending on the history of addition (or non-addition) of features seen so far. This leads naturally to the concept of considering a stream of features for model inclusion. Streamwise feature selection provides an adaptive method for controlling false positive, overcoming the problems of penalty methods which either assume fixed $m$ or ignore $m$.

3

We present two adaptive complexity penalty methods for streamwise feature se-lection, information-investing and $\alpha$-investing, which dynamically adjust the thresh-old on the error reduction required for adding a new feature. These two methods give false discovery rate style guarantees against overfitting. They differ from standard penalty methods such as AIC, BIC and RIC, which always drastically over- or under-fit in the limit of infinite numbers of non-predictive features. Information-investing is formulated in an minimum description length setting. The code length of data is decomposed into two parts: One part is for encoding the model, and the other part is for encoding the residual error given the model. $\alpha$-investing is formulated using a t or F statistic.

When the space of potential features is large, streamwise feature selection offers many advantages over traditional feature selection methods, which assume that all features are known in advance. Since most feature will not be included in the models, they can be discarded soon after generation, thus reducing data storage requirements. By eliminating the need to retain every feature, we can solve larger problems than can be tackled using standard machine learning algorithms such as support vector machines or stepwise regression which assume that all potential predictive features are known a priori.

Better models can often be constructed by generating new features based on the raw feature set. Transformation of raw features will usually give us more useful information. Useful transformations include log, square root, etc of single variables, and interactions. When selecting features using streamwise regression from poten-tially enormous sets of features, it is desirable to interleave the process of generating new features with that of feature testing. Features can be generated dynamically, focusing the search for new features on promising subspaces, and overfitting can be controlled by dynamically adjusting the threshold for adding features to the model.

Empirical results show that streamwise regression is competitive with (on small data sets) and superior to (on large data sets) much more compute-intensive feature selection methods such as stepwise regression, and allows feature selection on problems with millions of potential features.

## 1.2   Simultaneous Multiple Regressions using MIC

When doing feature selection in simultaneous multiple regressions, one can "borrow strength" across the different regressions to get a more sensitive criterion for which features to include in the models. A feature which is predictive for one response is more likely to be predictive for other responses. Suppose a feature is marginal useful for response A but significant useful for response B, and response A is correlated with B. Then response A might borrow strength from B to include this feature, and how much the strength A can borrow is determined by the correlation between A and B. Making use of such sharing of features should improve the average prediction accuracy of regressions.

This book extends streamwise feature selection to carry this out by using the two-part minimum description length principle. We call the derived feature selection method the multiple inclusion criterion (MIC). We use information theory to derive a model complexity penalty similar to the risk inflation criterion (RIC), which provides a minimax guarantee against overfitting. The code length for encoding the model is decomposed into three parts. Part one is for encoding the feature index, part two for encoding how many and which of the regressions include the feature, and part three for encoding the feature coefficients in the regressions.

In MIC, features are simultaneously considered for inclusion in all the regressions or in a subset of the regressions, and an efficient coding scheme is used to specify which features to include in which of the regression models. The MIC approach is useful both for predicting multiple responses[1] from the same set of potential features,

---

[1]We call "response" the response label column in the response matrix or the response variable in a regression.

and also for predicting similar responses from similar features, e.g. in trials on different populations, where different, but related features are used in different trials, but one still wishes to borrow strength across the different populations.

## 1.3 Applications of Feature Selection

In the past few years, more and more problems require analyzing big data sets. We focus on a particular problem: analyzing microarray data. Such kind of data sets usually have tens or hundreds of thousands of features but only several hundreds of observations. Here, the observations are persons, the response can be single binary variable indicating disease vs. non-disease or multiple binary variables indicating different cancers respectively, and each feature is a gene with continuous values of gene expression levels corresponding to the presence of mRNA. The classification task is selecting useful genes to predict which person has disease or which person has which cancers.

Another form of microarray data have genes as the observations. The response are multiple continuous variables indicating the gene expression levels under different conditions. Each response matrix data column representing a different condition, and the different conditions means different tissues or measurement environments. Each feature is a motif (transcription factor binding sit; it is a DNA sequence) with continuous values representing what strength of stickiness the site has. The regression task is finding the most significant motifs.

Another kind of data set arises in text mining. For example, people want to classify documents into different groups, the documents in each group having the common topic. The documents are represented by a "bag-of-words". That is, the vocabulary words are the features with values of word frequency counts of the observations or documents (proper normalization of the feature values also apply). Other applications include word sense disambiguation and part-of-speech tagging problems in which features are the words around the target word and the task is to identify the sense or part-of-speech of the target word.

## 1.4 Book Outline

This book will proceed as follows. Chapter 2 presents some basic information theory results and introduces the two-part minimum description length principle which will be used in later chapters to derive streamwise feature inclusion criterion. Chapter 3 presents the streamwise feature selection framework and its two algorithms, information-investing and $\alpha$-investing. Chapter 4 explores simultaneous multiple regressions using stepwise and streamwise regressions. Chapter 5 concludes and proposes future work.

# Chapter 2

# Information Theory Background

## 2.1  Definitions

We would like to transmit a message to a receiver. The message is composed of symbols and each of these symbols is from the same alphabet. The message can have repeated appearance of symbols. In order to transmit the message, we need to encode it first. In this book, we assume we use binary codes to encode the message. This means that we will use a unique sequence of 0 or 1 to encode each symbol of the message. Such unique binary code sequence for a symbol is called "symbol codeword".

To encode the message, we can concatenate the symbol codewords according to the symbol order within the message. Such concatenated binary code sequence for a message is called "message codeword".

A good message encoding scheme must have two kinds of one-to-one mapping properties. The first one-to-one mapping property requires that a symbol has a unique symbol codeword and a symbol codeword represents a unique symbol. The second one-to-one mapping property requires that a message has a unique message codeword and a message codeword represents a unique message. That is, in order to decode a message codeword, the receiver need to know how to break the message codeword into a unique sequence of symbol codewords. If a message codeword has multiple breaking options, the receiver will be confused when he want to recover the message from the message codeword. For example, if "0" represents "a", "1" represents "b", and "01" represents "c". Then, if facing a message codeword "011", the receiver don't know if he should decode it as "abb" or "cb". The fact that the symbol codeword "0" of "a" is a prefix of the symbol codeword "01" of "c" results in this problem.

## 2.2 Prefix Code

The sender can avoid the confusion by using an encoding scheme with prefix property. The prefix property states that no symbol codeword appears as a prefix of the other symbol codewords. A code with the prefix property is called "prefix code". For example, continue our example: if we use "00" represents "a", then the encoding scheme has prefix property. "011" will be decoded uniquely as the message "cb".

If we see the encoding scheme as a tree, every node of the tree represents a possible symbol codeword. Designing an encoding scheme is equivalent to assigning the nodes (codewords) to each symbol of the alphabet. If one node is selected as a symbol codeword, its children will not be selected as codewords for other symbols in order to fulfill the prefix property requirement.

## 2.3 Kraft Inequality

If an encoding scheme possesses the prefix property, it will satisfy kraft inequality

$$\sum_{\tau \in A} 2^{-L(\tau)} \leq 1,$$

where $A$ is the alphabet and $\tau$ is the symbol belonging to the alphabet. Note that if an encoding scheme satisfy the kraft inequality, then there always exits a prefix code for this scheme.

9

If the encoding scheme is efficient, the "≤" will be replaced by "=" and the scheme is "kraft tight". If partial bit is allowed and an encoding scheme satisfies kraft inequality but is not kraft tight, then there should be at least one symbol we can use less (partial) bits to encode than the current encoding scheme specifies. The term $2^{-L(\tau)}$ is non-negative and, if an encoding scheme is kraft tight, summation of the terms over all symbols is equal to one. Therefore, this term behaves like a probability function: higher frequency the symbol has, less bits we will use to encode the symbol.

## 2.4 Optimal Code Length

We always want to use less bits to encode messages. In the tree, if we have selected a set of nodes (symbol codewords) for assigning to symbols (assume the selection satisfy prefix property and hence also satisfy the kraft inequality), we then need to decide which node (symbol codeword) is assigned to which symbol. To use less bits to code messages, we will aim to minimize the expected mean length of the message codewords. Since a message codeword is composed of symbol codewords, the purpose can be achieved by minimizing the expected mean length of symbol codeword:

$$E(L) = \sum_{\tau \in A} P(\tau)L(\tau)$$

where $P(\tau)$ is the probability of occurrences of symbol $\tau$, and $L(\tau)$ is the length of $\tau$'s symbol codeword.

Shannon showed that the "ideal" optimal expected code length $E(L)$ for symbols is achieved by assigning $L(\tau) = -\log P(\tau)$ bits to encode the symbol $\tau$. Then we have below optimal expected code length:

$$E(L) = -\sum_{\tau \in A} P(\tau)\log P(\tau)$$

This equation relates code length with entropy under an assumption of probability $P$. That means, if we want to know how many bits in expectation we need to encode data, we only need to calculate the data's entropy. This gives us one-to-one relationship between code length and data's probability function.

The entropy represents a bound of the code length.

Shannon's Source Coding Theorem: Suppose symbols of $A$ are generated according to a probability distribution $P$. For any prefix code $C$, the expected mean length of the symbol codeword is equal to or greater than the entropy of $P$.

$$L_c \geq - \sum_{\tau \in A} P(\tau) \log P(\tau)$$

The intuition behind: we can economize transmission time by assigning short codewords to common symbols and long codewords to rare symbols.

## 2.5 Code Length for Integers

Code length for finite integers: consider a finite collection of integers $A = 1, 2, 3, ..., N$. If very little is known about how the data were generated, we can assume a uniform distribution. The uniform distribution assigns probabilities $1/N$ to each element of $A$. Then, for each element $\tau$, the resulting code length by applying Huffman's algorithm is

$$L(\tau) = \lfloor \log(N) \rfloor$$

where $\lfloor \log(N) \rfloor$ is the integer part of $\log(N)$.

When $\tau$ is a large positive integer, Elias [12] argues that if the binary representation is followed by other binary data, in addition to the basic $\log(N)$ bits, further bits are need to encode where the binary representation of $N$ ends. These additional bits are called the "preamble". It can be shown that the resulting code length is

$$L(\tau) = \log^*(\tau) \equiv \log c + \log \tau + \log \log \tau + ...$$

where $c \approx 2.865$. Since only positive item can be included, above definition can be written as:

$$L(\tau) = \log^*(\tau) \equiv \log c + \sum_{j>1} \max(\log^{(j)} \tau, 0)$$

11

where $\log^{(j)}(.)$ is the $j$th composition of log, such as, $\log^{(3)} \tau = \log \log \log \tau$.

Rissanen shows that if $P(\tau)$ is the probability function of $\tau$, then

$$\lim_{N \to \infty} \frac{- \sum_{\tau=1}^{N} P(\tau) \log^*(\tau)}{- \sum_{\tau=1}^{N} P(\tau) \log P(\tau)} = 1$$

This means that if $\tau$ is large enough, the $\log^*(\tau)$ is the optimal code length no matter what $p(\tau)$ is. Since $\log^*(.)$ satisfy the prefix property, $2^{-\log^*(\tau)}$ is called a "universal prior" for positive integer $\tau$.

## 2.6 Code Length for Real Numbers

Since it will take infinite bits to encode a real number $\tau$, we need to truncate the real number to a given precision $\delta$ so that

$$|\tau - \tau_\delta| < \delta$$

where $\tau_\delta$ is the truncated real number. With a careful chosen precision $\delta$, we can instead encode the truncated number $\tau_\delta$ without losing much information of $\tau$. Since

$$\tau_\delta = \lfloor \tau_\delta \rfloor + frac(\tau_\delta)$$

where $\lfloor \tau_\delta \rfloor$ is the integer part of $\tau_\delta$ and $frac(\tau_\delta)$ is the fractional part of $\tau_\delta$, we need to encode the two part respectively. Therefore, when $\tau$ is large,

$$L(\tau_\delta) = \log^*(\lfloor \tau_\delta \rfloor) + \log(1/\delta) \approx \log \tau - \log \delta$$

If the density function of $\tau$ is $f(\tau)$, then we have

$$L(\tau_\delta) = - \log f(\tau) - \log \delta$$

## 2.7 Minimum Description Length

Let us suppose data $y$ (for example, the response vector in regression analysis) is a symbol in $A$, $y$ is generated from a model $f$, and the sender wants to encode $y$ and send the symbol message to the receiver so that the receiver can decode and recover the value of $y$ from the symbol message. As we discussed before, to get the minimum mean code length of $y$, the sender can encode $y$ such that the code length is $-\log f(y)$. If both sides knows what the model $f$ is (including its parameter estimate $\hat{\theta}$), we can use less bits to encode $y$ by just coding the residual error given the model, and the code length is as:

$$L(y) = -\log f(y|\hat{\theta}).$$

We get the estimate $\hat{\theta}$ of $\theta$ by minimizing $-\log f(y|\theta)$ or maximizing the data likelihood given the model, that is,

$$L(y) = -\log f(y|\hat{\theta}) = \min_{\theta}\{-\log f(y|\theta)\}.$$

In this sense, MDL is equivalent to the maximum likelihood method.

If the sender and receiver only know $f$ is a member of a class of models:

$$M = \{f(y|\theta) : \theta \in \Theta \subset R^k\}$$

where $\Theta$ is the parameter space. The sender will send a code message to notify the receiver which member of the model class $M$ generates data $y$. Therefore, the code message will consist of two parts: one part is the code for describing the model and the other part is the code for describing the residual error given the model. It is the principle of minimum description length. Let us state it as:

$$L(\text{data}) = L(\text{model}) + L(\text{residual error given the model}).$$

For example, we consider feature selection in linear regression:

$$y = \sum_{\gamma_j=1} \beta_j x_j + \varepsilon$$

13

where $\varepsilon \sim N(0, \sigma^2)$ (Gaussian distribution with mean zero and unknown variance $\sigma^2$) and $\gamma_j$ is a indicator: if $\gamma_j = 1$, then $j$th feature is included into the model; if $\gamma_j = 0$, then $j$th feature is not included into the model, that is,

$$\gamma = (\gamma_1, \gamma_2, ..., \gamma_m) \in 0, 1^m$$

when we have $m$ candidate features. Therefore, there are $2^m$ possible models. The vector $\gamma = (\gamma_1, ..., \gamma_m)$ gives us the index of candidate models. For simplicity, let $\hat{\theta}_\gamma$ denotes the coefficients of the features which are included into the model, that is, $\hat{\theta}_\gamma \equiv \hat{\theta}_{\{\gamma_j = 1\}}$. Then, the code length required for encoding the data $y$ is:

$$L(y) = L(\gamma, \hat{\theta}_\gamma) + L(y|\gamma, \hat{\theta}_\gamma)$$

where $L(\gamma, \hat{\theta}_\gamma)$ is the cost (in bits) to encode the model which is indexed by $\gamma$ and have parameter $\hat{\theta}_\gamma$, and $L(y|\gamma, \hat{\theta}_\gamma)$ is the cost (in bits) to encode the residual error given the model.

We assume that each $\gamma_i$ is a Bernoulli random variable with $p = 0.5$. Then for each outcome of $\gamma$ (that is, for each member of the model class), we have the probability:

$$P(\gamma) = (1/2)^q (1/2)^{m-q} = (1/2)^m$$

Then the ideal code length for encoding this outcome of $\gamma$ is:

$$L(\gamma) = -\log P(\gamma) = m$$

That is, it costs $m$ bits to indicate which features are included in the model. Each of the $m$ candidate feature is assigned one bit. If the feature is included, the bit is one; if not included, the bit is zero. This coding scheme assumes that each feature has equal chance to enter into the model. We know it is not true in the real data sets. We developed algorithm to estimate the probability of entering model for each feature and use the probability to encode the model's feature index efficiently.

As we discussed above, if we know the probability density function of each included feature, we can encode the parameter vector $\hat{\theta}_\gamma$ as:

$$L(\hat{\theta}_\gamma) = -\sum_{j=1}^{q} \log f(\hat{\theta}_j) - \sum_{j=1}^{q} \log \delta_j$$

when there are $q$ features added into the model. Therefore, the code length for encoding the model is:

$$L(\gamma, \hat{\theta}_\gamma) = m - \sum_{j=1}^{q} \log f(\hat{\theta}_j) - \sum_{j=1}^{q} \log \delta_j$$

Given the model or density function $f$, the log-likelihood of data $y$ specify the code length of the second part of total description length:

$$L(y|\gamma, \hat{\theta}_\gamma) = -\log f(y|\gamma, \hat{\theta}_\gamma)$$

Therefore, we have the total description length (when there are $q$ features in the model):

$$L(y) = m - \sum_{j=1}^{q} \log f(\hat{\theta}_j) - \sum_{j=1}^{q} \log \delta_j - \log f(y|\gamma, \hat{\theta}_\gamma)$$

When sample size $n$ is large enough, $-\sum_{j=1}^{q} \log \delta_j = \frac{q}{2} \log(n)$, and $-\sum_{j=1}^{q} \log f(\hat{\theta}_j)$ can be ignored because $\frac{q}{2} \log(n)$ is the dominating term. This will result in a description length of

$$L(y) = \frac{q}{2} \log(n) - \log f(y|\gamma, \hat{\theta}_\gamma) \tag{2.1}$$

This description length formulation is equivalent to BIC method in feature selection. BIC is valid coding scheme if the sample size is large enough. But when the feature size is much larger than the sample size, that is, $m \gg n$, BIC will overfit severely, selecting many spurious features. We will discuss it later.

Equation 2.1 gives us the sense of the general form of minimum description length. There is a trade-off problem in feature selection: complex model will give us small residual error (in-sample error) and small code length for encoding it but large code

15

length for encoding the model. The number of bits for encoding the model is like a penalty of model complexity, so MDL resembles the penalized likelihood method in feature selection, that is, we can simplify MDL formulation within the model or feature selection framework:

$$MDL \equiv \min_{f} \{ -\log(\text{ likelihood } | f) + PE(f) \} \qquad (2.2)$$

where $PE(f)$ represents the penalty function designed to penalize model complexity and its value is determined by the number of features included in the model, that is, $q$.

# Chapter 3

# Streamwise Feature Selection

## 3.1 Introduction

In many predictive modeling tasks, one has a fixed set of observations from which a vast, or even infinite, set of potentially predictive features can be computed. Of these features, often only a small number are expected to be useful in a predictive model. Pairwise interactions and data transformations of an original set of features are frequently important in obtaining superior statistical models, but expand the number of feature candidates while leaving the number of observations constant. For example, in a recent bankruptcy prediction study [17], pairwise interactions between the 365 original candidate features led to a set of over $67,000$ resultant candidate features, of which about 40 proved to be significant. The feature selection problem is to identify and include features from a candidate set with the goal of building a statistical model with minimal out-of-sample (test) error. As the set of potentially predictive features becomes ever larger, careful feature selection to avoid overfitting and to reduce computation time becomes ever more critical.

In this chapter, we describe *streamwise feature selection*, a class of feature selection methods in which features are considered sequentially for addition to a model, and either added to the model or discarded, and two simple streamwise regression algorithms[1], information-investing and $\alpha$-investing, that exploit the streamwise feature setting to produce simple, accurate models. Figure 3.1 gives the basic framework of streamwise feature selection. One starts with a fixed set of $y$ values (for example, labels for observations), and each potential feature is sequentially tested for addition to a model. The threshold on the required benefit (for example, error or entropy reduction, or statistical significance) for adding new features is dynamically adjusted in order to optimally control overfitting.

Streamwise regression should be contrasted with "batch" methods such as *stepwise* regression or support vector machines (SVMs). In stepwise regression, there is no order on the features; all features must be known in advance, since all features are evaluated at each iteration and the best feature is added to the model. Similarly, in SVMs or neural networks, all features must be known in advance. (Overfitting in these cases is usually avoided by regularization, which leaves all features in the model, but shrinks the weights towards zero.) In contrast, in streamwise regression, since potential features are tested one by one, they can be generated dynamically.

By modeling the candidate feature set as a dynamically generated stream, we can handle candidate feature sets of unknown, or even infinite size, since not all potential features need to be generated and tested. Enabling selection from a set of features of unknown size is useful in many settings. For example, in statistical relational learning [27, 11, 10], an agent may search over the space of SQL queries to augment the base set of candidate features found in the tables of a relational database. The number of candidate features generated by such a method is limited by the amount of CPU time available to run SQL queries. Generating 100,000 features can easily take 24 CPU hours [38], while millions of features may be irrelevant due

---

[1]The algorithms select features and add these features into regression models. Since feature selection and regression are closely coupled here, we use "streamwise feature selection" and "streamwise regression" interchangeably. Some papers use the terms "regression" for continuous responses and "classification" for categorical responses. We use "regression" for both cases, since generalized linear regression methods such as logistic regression handle categorical responses well.

**Input:** A vector of $y$ values (for example, labels), and a stream of features $x$.

{initialize}

model = {}                              //initially no features in model

$i = 1$                                 // index of features

**while** CPU_time_used < max_CPU_time **do**

    $x_i \leftarrow$ get_next_feature()

    {Is $x_i$ a "good" feature?}

    **if** fit_of($x_i$, model) > threshold **then**

        model $\leftarrow$ model $\cup$ $x_i$  // add $x_i$ to the model

        decrease threshold

    **else**

        increase threshold

    **end if**

    $i \leftarrow i + 1$

**end while**

Figure 3.1: Algorithm: general framework of streamwise feature selection. The threshold on statistical significance of a future new feature (or the entropy reduction required for adding the future new feature) is adjusted based on whether current feature was added. fit_of($x_i$, model) represents a score, indicating how much adding $x_i$ to the model improves the model. Details are provided below.

to the large numbers of individual words in text. Another example occurs in the generation of transformations of features already included in the model (for example, pairwise or cubic interactions). When there are millions or billions of potential features, just generating the entire set of features (for example, cubic interactions or three-way table merges in SQL) is often intractable. Traditional regularization and feature selection settings assume that all features are pre-computed and presented to a learner before *any* feature selection begins. Streamwise regression does not.

Streamwise feature selection can be used with a wide variety of models where p-values or similar measures of feature significance are generated. We evaluate streamwise regression using linear and logistic regression (also known as maximum entropy modeling), where a large variety of selection criteria have been developed and tested. Although streamwise regression is designed for settings in which there is some prior knowledge about the structure of the space of potential features, and the feature set

size is unknown, in order to compare it with stepwise regression, we apply streamwise regression in traditional feature selection settings, that is, those of fixed feature
set size. In such settings, empirical evaluation shows that, as predicted by theory,
for smaller feature sets such as occur in the UCI data sets, streamwise regression
produces performance competitive to stepwise regression using traditional feature
selection penalty criteria including AIC [2], BIC [41], and RIC [9, 14]. As feature set
size becomes larger, streamwise regression offers significant computational savings
and higher prediction accuracy.

The ability to do feature selection well encourages the use of different transformations of the original features. For sparse data, principal components analysis
(PCA) or other feature extraction methods generate new features which are often
predictive. Since the number of potentially useful principal components is low, it
costs very little to generate a couple different projections of the data, and to place
these at the head of the feature stream. Smaller feature sets should be put first.
For example, first PCA components, then the original features, and then interaction
terms. Results presented below confirm the efficiency of this approach.

Features in the feature stream can be sorted by cost. If features which are cheaper
to collect are placed early in the feature stream, they will be preferentially selected
over redundant expensive features later in the stream. When using the resulting
model for future predictions, one needs not collect the redundant expensive features.

Alternatively, features can be sorted so as to place potentially higher signal con-
tent features earlier in the feature stream, making it easier to discover the useful
features. Different applications benefit from different sorting criteria. For example, sorting gene expression data on the variance of features sometimes helps (see
Section 3.6.2). Often features come in different types (person, place, organization;
noun, verb, adjective; car, boat, plane). A combination of domain knowledge and
use of the different sizes of the feature sets can be used to provide a partial order
on the features, and thus to take full advantage of streamwise feature selection. As

20

described below, one can also dynamically re-order feature streams based on which features have been selected so far.

## 3.2   Traditional Feature Selection: A Brief Review

Traditional feature selection typically assumes a setting consisting of $n$ observations and a fixed number $m$ of candidate features. The goal is to select the feature subset that will ultimately lead to the best performing predictive model. The size of the search space is therefore $2^m$, and identifying the best subset is NP-complete. Many commercial statistical packages offer variants of a greedy method, stepwise feature selection, an iterative procedure in which at each step all features are tested at each iteration, and the single best feature is selected and added to the model. Stepwise regression thus performs hill climbing in the space of feature subsets. Stepwise selection is terminated when either all candidate features have been added, or none of the remaining features lead to increased expected benefit according to some measure, such as a p-value threshold. We show below that an even greedier search, in which each feature is considered only once (rather than at every step) gives competitive performance. Variants of stepwise selection abound, including forward (adding features deemed helpful), backward (removing features no longer deemed helpful), and mixed methods (alternating between forward and backward). Our evaluation and discussion will assume a simple forward search.

There are many methods for assessing the benefit of adding a feature. Computer scientists tend to use cross-validation, where the training set is divided into several (say $k$) batches with equal sizes. $k - 1$ of the batches are used for training while the remainder batch is used for evaluation. The training procedure is run $k$ times so that the model is evaluated once on each of the batches and performance is averaged. The approach is computationally expensive, requiring $k$ separate retraining steps for each evaluation. A second disadvantage is that when observations are scarce the

21

method does not make good use of the observations. Finally, when many different models are being considered (for example, different combinations of features), there is a serious danger of overfitting when cross-validation is used. One, in effect, is selecting the model to fit the test set.

Penalized likelihood ratio methods [5] for feature selection are preferred to cross-validation by many statisticians, as they do not require multiple re-trainings of the model and they have attractive theoretical properties. Penalized likelihood can be represented as:

$$\text{score} = -2\log\left(\text{likelihood}\right) + F \times q \qquad (3.1)$$

where $F$ is a function designed to penalize model complexity, and $q$ represents the number of features currently included in the model at a given point. The first term in the equation represents a measure of the in-sample error given the model, while the second is a model complexity penalty. Table 3.1 contains the definitions which we use throughout this chapter. In addition, we define *beneficial*[2] or *spurious* features as those which, if added to the current model, would or would not reduce prediction error, respectively, on a hypothetical infinite large test data set. Note that under this definition of beneficial, if two features are perfectly correlated, the first one in the stream would be beneficial and the second one spurious, as it would not improve prediction accuracy. Also, if a prediction requires an exact XOR of two features, the raw features themselves could be spurious, even though the derived XOR-feature might be beneficial. We speak of the *set of beneficial features in a stream* as those which would have improved the prediction accuracy of the model at the time they were considered for addition if all prior beneficial features had been added.

Only features that decrease the score defined in equation 4.1 are added to the model. In other words, the benefit of adding the feature to the model as measured

---

[2]Some papers use the terms "useful" or "relevant"; please see [30] and [6] for a discussion and definitions of these terms. If the features were independent (orthogonal), then we could speak of "true" features, which improve prediction accuracy for a given classification method regardless of which other features are already in the model.

| Symbol | Meaning |
|--------|---------|
| $n$ | Number of observations |
| $m$ | Number of candidate features |
| $m^*$ | Number of beneficial features in the candidate feature set |
| $q$ | Number of features currently included in a model |

Table 3.1: Symbols used throughout this chapter and their definitions.

| Name | Nickname | Penalty |
|------|----------|---------|
| Akaike information criterion | AIC | 2 |
| Bayesian information criterion | BIC | $\log(n)$ |
| risk inflation criterion | RIC | $2\log(m)$ |

Table 3.2: Different choices for the model complexity penalty $F$.

by the likelihood ratio must surpass the penalty incurred by increasing the model complexity. We focus now on choice of $F$. Many different functions $F$ have been used, defining different criteria for feature selection. The most widely used of these criteria are the Akaike information criterion (AIC), the Bayesian information criterion (BIC), and the risk inflation criterion (RIC). Table 3.2 summarizes the penalties $F$ used in these methods.

For exposition we find it useful to compare the different choices of $F$ as alternative coding schemes for use in a minimum description length (MDL) criterion framework [40]. In MDL, both sender and receiver are assumed to know the feature matrix and the sender wants to send a coded version of a statistical model and the residual error given the model so that the receiver can construct the response values. Equation 4.1 can be viewed as the length of a message encoding a statistical model (the second term in equation 4.1) plus the residual error given that model (the first term in equation 4.1). To encode a statistical model, an encoding scheme must identify which features are selected for inclusion and encode the estimated coefficients of the included features. Using the fact that the log-likelihood of the data given a model gives the number of bits to code the model residual error leads to the criteria for feature selection: accept a new feature $x_i$ only if the change in log-likelihood from adding the feature is greater than the penalty $F$, that is, if $2\log(P(y|\hat{y}_{x_i})) -$

$2 \log(P(y|\hat{y}_{-x_i})) > F$ where $y$ is the response values, $\hat{y}_{x_i}$ is the prediction when the feature $x_i$ is added into the model, and $\hat{y}_{-x_i}$ is the prediction when $x_i$ is not added. Different choices for $F$ correspond to different coding schemes for the model.

Better coding schemes encode the model more efficiently; they produce a more accurate depiction of the model using fewer bits. AIC's choice of $F = 2$ corresponds to a version of MDL which uses *universal priors* for the coefficient of a feature which is added into the model [15]. BIC's choice of $F = \log(n)$ employs more bits to encode the coefficient as the training set size grows larger. Using BIC, each zero coefficient (feature not included in the model) is coded with one bit, and each non-zero coefficient (feature included in the model) is coded with $1 + \frac{1}{2} \log(n)$ bits (all logs are base 2). BIC is equivalent to an MDL criterion which uses *spike-and-slab priors* if the number of observations $n$ is large enough [42].

However, neither AIC nor BIC are valid codes for $m \gg n$. They thus are expected to perform poorly as $m$ grows larger than $n$, a situation common in streamwise regression settings. We confirm this theory through empirical investigation in Section 3.6.2.

RIC corresponds to a penalty of $F = 2 \log(m)$ [14, 21]. Although the criterion is motivated by a minimax argument, following [42] we can view RIC as an encoding scheme where $\log(m)$ bits encode the index of which feature is added. Using RIC, no bits are used to code the coefficients of the features that are added. This is based on the assumption that $m$ is large, so that the $\log(m)$ cost dominates the cost of specifying the coefficients. Such an encoding is most efficient when we expect few of the $m$ candidate features enter the model.

RIC can be problematic for streamwise feature selection since RIC requires that we know $m$ in advance, which is often not the case (see Section 3.3). We are forced to guess a $m$, and when our guess is inaccurate, the method may be too stringent or not stringent enough. By substituting $F = 2 \log(m)$ into Equation 4.1 and examining the resulting chi-squared hypothesis test, it can be shown that the p-value required to reject the null hypothesis must be smaller than $\frac{0.05}{m}$. In other words, RIC may be viewed as a Bonferroni p-value thresholding method. Bonferroni methods are known to be overly stringent [4], a problem exacerbated in streamwise feature selection applications when $m$ should technically be chosen to be the largest number of features that might be examined. On the other hand, if $m$ is picked to be a lower bound of the number of predictors that might be examined, then it is too small and there is increased risk that some feature will appear by chance to give significant performance improvement.

Streamwise feature selection is closer in spirit to an alternate class of feature selection methods that control the false discovery rate (FDR), the fraction of the features that are added to the model that reduce predictive accuracy (Benjamini and Hochberg, 1995). Unlike AIC, BIC and RIC, which require each potential feature to be above the same threshold, FDR methods compute p-values (here, the probability of feature increasing test error), sort the features by p-value, and then use a threshold which depends on both the total number of features considered (like RIC) and the number of features that have been added, making use of the fact that adding some features which are almost certain to reduce prediction error allows us to add other features which are more marginal, while still meeting the FDR criterion. In this chapter we propose an alternative to FDR that, among other benefits, can handle infinite feature streams, and make the above claims precise.

## 3.3  Interleaving Feature Generation and Testing

In streamwise feature selection, candidate features are sequentially presented to the modeling code for potential inclusion in the model. As each feature is presented, a decision is made using an adaptive penalty scheme as to whether or not to include the feature in the model. Each feature needs be examined at most once.

The "streamwise" view supports flexible ordering on the generation and testing of features. Features can be generated dynamically based on which features have already been added to the model.[3] Note that the theory provided below is independent of the feature generation scheme used. All that is required is a method of generating features, and an estimation package which given a proposed feature for addition to the model returns a p-value for the corresponding coefficient or, more generally, the change in likelihood of the model resulting from adding the feature. One can also test the same feature more than once (as in stepwise regression), but we have not found significant benefit from doing multiple passes through the features.

New features can be generated in many ways. For example, in addition to the $m$ original features, $m^2$ pairwise interaction terms can be formed by multiplying all $m^2$ pairs of features together. (Almost half of these features are, of course, redundant with the other half due to symmetry, and so need not be generated and tested.) We refer to the interaction terms as *generated* features; they are examples of a more general class of features formed from transformations of the original features (square root, log, *etc.*), or combinations of them including, for instance, PCA. Such strategies are frequently successful in obtaining better predictive models.

Rather than testing all possible interactions in an arbitrary order, it is generally better to initially test interactions of the features that have already been selected with themselves, then to test interactions of the selected features with the original features, and finally (if computer power permits) to test all interactions of the original features. This requires dynamic generation of the feature stream, since the first interaction terms can not be specified in advance, as they depend on which features have already been selected. (It can be the case, as in an XOR or parity problem, that interactions are significant when none of the individual component features are, but it still makes sense as a search strategy to try the smaller parts of the feature space first.)

---

[3] One cannot use the coefficients of the features that were not added to the model, since stream-wise regression does not include the cost of coding these coefficients, and so this *would* lead to overfitting. One can, of course, use the rejected features themselves in interaction terms, just not their coefficients.

Statistical relational learning (SRL) methods can easily generate millions of potentially predictive features as they "crawl" through a database or other relational structure and generate features by building increasingly complex compound relations or SQL queries [38]. For example, when building a model to predict the journal in which an article will be published, potentially predictive features include the words in the target article itself, the words in the articles cited by the target article, the words in articles that cite articles written by the authors of the target article, and so forth.

Both stepwise regression and standard shrinkage methods require knowing all features in advance, and are poorly suited for the feature sets generated by SRL. Since stepwise regression tests all features for inclusion at each iteration, it is computational infeasible on large data sets. Even if computer speed and memory were not an issue, control of overfitting using standard penalty methods would fail. Some other strategy such as streamwise feature selection is required. Interleaving the generation of features with the assessment of model improvement allows the search over potential features to be pruned to promising regions. A potentially intractable search thus becomes tractable.

In SRL, one searches further in those branches of a refinement graph where more component terms have proven predictive. In searching for interaction terms, one looks first for interactions or transformations of features which have proven significant. This saves the computation, and more importantly, avoids the need to take a complexity penalty for the many interaction terms which are never examined.

There are also simple ways to dynamically interleave multiple kinds of features, each of which is in its own stream. The main feature stream used in streamwise regression is dynamically constructed by taking the next feature from the sub stream which has had the highest success in having features accepted. If a previously successful stream goes long enough without having a feature selected, then other streams will be automatically tried. To assure that all streams are eventually tried, one can use a score for each stream defined as (number of features selected + $a$)/(number of features tried + $b$). The exact values of $a$ and $b$ do not matter, as long as both are positive. A single feature stream is used in this chapter.

## 3.4 Streamwise Regression using Information-investing

Streamwise regression can be used either in an MDL setting ("information-investing") or in a statistical setting using a t or F statistic ("$\alpha$-investing"). We first present streamwise regression in an information-investing setting. Information-investing [44] is derived using a minimum description length (MDL) approach [42]. From a coding viewpoint, we wish to transmit a message to a receiver in order to let the receiver get the response values ($y$), assuming that the receiver knows $x$. In this sense, the score in equation 4.1 is the description length required to code this message. The model is then chosen that minimizes the description length. If a feature is added to the model and reduces the description length, we call this reduction the *bits_saved*. Therefore, *bits_saved* is the decrease in the bits required to code the model error minus the increase in the bits required to code the model. The coding used to calculate *bits_saved* is described in details in Section 3.4.3. If *bits_saved* is larger than a threshold, we add the feature to the model. The algorithm is shown in Figure 3.2. We set both $W_0$ and $W_\Delta$ to 0.5 bit in all of the experiments presented in this chapter.

Information-investing allows us, for any valid coding, to have a false discovery rate (FDR) style bound, and thus to minimize the expected test error by adding as many features as possible subject to controlling the FDR bound [46].

Streamwise regression with information-investing consists of three components:

- *Wealth Updating*: a method for adaptively adjusting the bits available to code the features which will not be added to the model.

**Input:** A vector of $y$ values (for example, labels), a stream of features $x$, $W_0$, and $W_\Delta$.

{initialize}
model = {}                      //initially no features in model
$i = 1$                         // index of features
$w_1 = W_0$                     // initial bits available for coding
**while** CPU_time_used < max_CPU_time **do**
    $x_i \leftarrow$ get_next_feature()
    $\epsilon_i \leftarrow w_i/2i$                   // select bid amount
    {see Section 3.4.3 for the calculation of $bits\_saved$}
    **if** $bits\_saved(x_i, \epsilon_i, model) > W_\Delta$ **then**
        model $\leftarrow$ model $\cup\ x_i$ // add $x_i$ to the model
        $w_{i+1} \leftarrow w_i + W_\Delta$     // increase wealth
    **else**
        $w_{i+1} \leftarrow w_i - \epsilon_i$        // reduce wealth
    **end if**
    $i \leftarrow i + 1$
**end while**

Figure 3.2: Algorithm: streamwise regression using information-investing.

- *Bid Selection*: a method for determining how many bits, $\epsilon_i$, one is willing to spend to code the fact of not adding a feature $x_i$. Asymptotically, it is also the probability of adding this feature. We show below how bid selection can be done optimally by keeping track of the bits available to cover future overfitting (that is, the wealth).

- *Feature Coding*: a coding method for determining how many bits are required to code a feature for addition. We use a two part code, coding the presence or absence of the features, and then, if the feature is present, coding the sign and size of the estimated coefficient.

## 3.4.1  Wealth Updating

The information-investing coding scheme is adjusted using the wealth, $w$, which represents the number of bits currently available for future overfitting. The wealth is "invested" in testing features. Wealth starts at an initial value $W_0$.

Each time a feature is added, it is (in expectation) likely to be a beneficial feature and lead to a decrease in the total description length, leaving more bits available to risk future overfitting. Thus, wealth is increased by $W_\Delta$. By increasing wealth, we gain more feature selection power under the FDR bound. Our algorithm guarantees that the sum of wealth (which is increased by $W_\Delta$) and total description length (which is decreased by more than $W_\Delta$) is decreased. If a feature is not added to the model, $\epsilon$ bits is "invested" to code this fact and subtracted from wealth.

Streamwise feature selection allows us, for any valid coding, to bound in expectation the ratio of spurious features added to beneficial features added, and thus to minimize the expected test error by adding as many features as possible subject to controlling that ratio.

We first formally define *beneficial* features. For a classification function $f$ (for example, logistic regression) and a given set of features indexed as $\{j\}$ (for example, $\{1, 27, 192\}$ for the 1st, 27th, and 192th features), we want to know whether, in expectation over possible training sets $(y; X_{\{j\}})$ with the same set of features, adding a new feature with index $j_{new}$ decreases the expected out-of-sample (test set) error.

More formally, a feature indexed by $j_{new}$ is *beneficial* if adding it to a set of features $\{j\}$ reduces the expected out-of-sample error for classifier $f$. That is, if for a loss function $l(f, y)$, we have

$$E_{(y;X)} l\left(f\left(y; X_{\{j\}}\right), y\right) > E_{(y;X)} l\left(f\left(y; X_{\{j\}+j_{new}}\right), y\right)$$

where the expectation is over different training sets $(y; X)$, all for the same set of features $\{j\}$ or $\{j\} + j_{new}$. $\{j\} + j_{new}$ represents the new feature index set after a new feature indexed as $j_{new}$ is added.

**Theorem 1** *Let $S_i$ be the number of beneficial features included in the model, let $V_i$ be the number of spurious features included and $w_i$ be the wealth, all at iteration $i$, and let $W_\Delta \leq 1/6$ be a user selected value. Then if the algorithm in Figure 3.2 is* modified so that it never bids more than $1/3$ it will have the property that[4]:

$$E(V_i) \leq 2W_\Delta E(S_i) + 2W_0.$$

**Proof** The proof relies on the fact that $G_i \equiv V_i - 2W_\Delta S_i + 2w_i$ is a super-martingale [26], namely at each time period the conditional expectation of $G_i - G_{i-1}$ is negative. We will show that $G_i$ is a super-martingale by considering the cases when the feature is or is not a beneficial feature and is or is not added to the estimated model.

|  | spurious feature | beneficial feature |
|---|---|---|
| don't add the feature | $\Delta S_i = 0,\ \Delta V_i = 0$ | $\Delta S_i = 0,\ \Delta V_i = 0$ |
| add the feature | $\Delta S_i = 0,\ \Delta V_i = 1$ | $\Delta S_i = 1,\ \Delta V_i = 0$ |

We can write the change in $G_i$ as:

$$\begin{aligned} \Delta G_i &\equiv G_i - G_{i-1} \\ &= \Delta V_i - 2W_\Delta \Delta S_i + 2\Delta w_i. \end{aligned}$$

If the feature is a beneficial feature, then $\Delta V_i = 0$. Thus,

$$\Delta G_i = -2\epsilon_i(1 - \Delta S_i) \leq 0$$

where $\epsilon_i$ is the bid amount at time $i$. On the other hand, if the feature is a spurious feature, then $\Delta S_i = 0$. So,

$$\begin{aligned} \Delta G_i &= \Delta V_i + 2\Delta w_i \\ &= \Delta V_i + 2W_\Delta \Delta V_i - 2\epsilon_i(1 - \Delta V_i) \\ &= \Delta V_i(1 + 2W_\Delta + 2\epsilon_i) - 2\epsilon_i. \end{aligned}$$

We will show below that $\epsilon_i$ is approximately the probability of adding a spurious feature. Thus, $E(\Delta V_i) \leq \epsilon_i$. Also by assumption, $2W_\Delta \leq 1/3$ and $\epsilon_i \leq 1/3$.[5] Hence $E(\Delta G_i) \leq 2\epsilon_i - 2\epsilon_i = 0$. Thus, $G_i$ is a super-martingale.

---

[4] *The expectation $E(.)$ is over training sets drawn from any desired distribution. Given a hypothetical infinite test set, these training sets would be generated by random sampling.*

[5] We use $W_0 = 0.5$, corresponding to $\epsilon_1 = 0.25$, and $W_\Delta = 0.5$. This value of $W_\Delta$ does not satisfy the conditions of the theorem, but gives slightly better results in practice.

Using the weaker fact that for super-martingales: $E(G_i) \leq E(G_{i-1})$, we see that $E(G_i) \leq G_1$. But since we start out with $V_i = 0$ and $S_i = 0$, $G_1 = 2W_0$. Since $w_i > 0$ by construction, we see that:

$$E(V_i - 2W_\Delta S_i) \leq 2W_0.$$

■

When $W_\Delta = \frac{1}{6}$, this reduces to:

$$E(V_i) \leq \frac{1}{3}E(S_i) + 2W_0. \tag{3.2}$$

The expected number of spurious features added is thus no more than $2W_0$ greater than one third of the expected number of beneficial features added. This bound translates to a bound on the $mFDR = E(V)/E(S + V)$. mFDR [43] is a marginal version of the better known false discovery rate $FDR = E(V/S+V)$. Alternatively, one can use the excess discovery count: $EDC \equiv V - W_\Delta(S + V - 1)$ [see 18].

**Information-investing and out-of-sample error**

The information-investing, based on the principle of minimum description length, uses training data to build model. In fact, one might be more concerned with the out-of-sample total description length. Please see test results (Figure 3.3) on a synthetic data. The results show that in-sample total description length is always decreased when a feature is added into the model; although there might be a very slight corresponding increase in corresponding out-of-sample total description length, the trend of decrease of out-of-sample total description length is obvious. Since the true out-of-sample performance is evaluated on a hypothetical infinitely large test data set, the number of bits to encode a finite model is negligible and the out-of-sample total description length is equivalent to the out-of-sample error.

The relationship between the mFDR and minimizing out-of-sample error follows from the fact that the cost of adding a spurious feature is comparable to that of

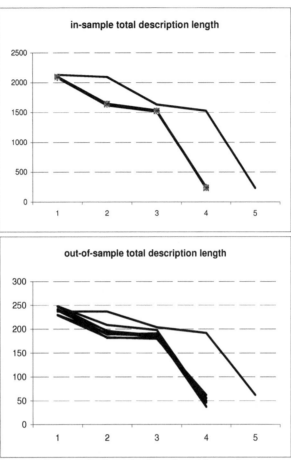

Figure 3.3: **In-sample total description length and corresponding out-sample totol description length.** Values on vertical axis are the total description length in bits and values on horizontal axis are the number of features in a model ($q$). The synthetic data has 900 observation in train set and 100 observations in test set, $m = 100$ features, and $m^* = 4$ beneficial features; $x \sim N(0,1)$, $y$ is linear in $x$ with noise $\sigma^2 = 0.1$. Beneficial features are randomly distributed in the feature set. 10 forward cross-validations (CVs); each line represents one CV. In the upper graph, nine lines (CVs) have so slight difference in values that they look like just one line (the below thicker line with markers); the fourth CV includes five features and its line departs away from the other nine lines.

failing to add a beneficial feature. Let $m^*$ be the total number of beneficial features in the candidate feature set, $S$ be the number of beneficial features added to the model, then $m^* - S$ be the number of beneficial features missed, $V$ be the number of spurious features added to the model, $\tilde{t}$ be the cutoff for the $t$-statistic, $\frac{\hat{\beta}}{\sqrt{n}\hat{\sigma}}$, of the estimated coefficient $\hat{\beta}$ to be significant, and $SE$ be the standard error in the model for $\hat{\beta}$. Define $\hat{\sigma}$ to be the in-sample estimate of the standard deviation of an out-of-sample forecast.

Adding a spurious feature will increase the out-of-sample sum of squared error by roughly $\hat{\sigma}^2 \hat{\beta}^2 / SE^2$. Failing to add a beneficial feature will increase the error by roughly $\hat{\sigma}^2 (\hat{\beta}^2 / SE^2 - 1)$.

A beneficial feature is difficult to find in that its $\hat{\beta}/SE$ is approximately $\tilde{t}$. We are interested in hard problems, namely ones that contain many such features. When a feature is spuriously included, its $t$-statistic will be close to $\tilde{t}$. Using these two facts, we can now estimate the out-of-sample sum of squared error as

$$(V\tilde{t}^2 + (m^* - S)(\tilde{t}^2 - 1))\hat{\sigma}^2.$$

We want to minimize the above error. Unfortunately, we cannot increase the number of beneficial features $S$ without also increasing the number of spurious features $V$ added. However, since incrementing $V$ by 1 increases $S$ by at least 3 (see Inequality 3.2), the error is changed by $(\tilde{t}^2 - 3(\tilde{t}^2 - 1))\hat{\sigma}^2 < 0$. Thus, if we were to change the feature acceptance criterion to either add more or add fewer features than suggested by the mFDR bound, then the out-of-sample fit would be expected to get worse. It is in this weak sense that our algorithm optimizes out-of-sample error.

### 3.4.2 Bid Selection

The selection of $\epsilon_i$ as $w_i/2i$ gives the slowest possible decrease in wealth such that all wealth is used; that is, so that as many features as possible are included in the model without systematically overfitting.[6]

**Theorem 2** *Computing $\epsilon_i$ as proportional to $w_i/2i$ gives the slowest possible decrease in wealth such that $\lim_{i \to \infty} w_i = 0$.*

**Proof** Define $\delta_i = \epsilon_i/w_i$ to be the fraction of wealth invested at time $i$. If no features are added to the model, wealth at time $i$ is $w_i = \Pi_i(1 - \delta_i)$. If we pass to the limit to generate $w_\infty$, we have $w_\infty = \Pi_i(1 - \delta_i) = e^{\sum \log(1-\delta_i)} = e^{-\sum \delta_i + O(\delta_i^2)}$. Thus, $w_\infty = 0$ iff $\sum \delta_i$ is infinite.

Thus if we let $\delta_i$ go to zero faster than $1/i$, say $i^{-1-\gamma}$ where $\gamma > 0$ then $w_\infty > 0$ and we have wealth that we never use. ∎

### 3.4.3 Feature Coding

To code an added feature, we code both the fact that the feature is added and the value of its estimated coefficient. Since $\epsilon$ is the number of bits available to code the fact of not adding a feature, the probability of not adding a feature should be $e^{-\epsilon}$ if the coding is optimal. Therefore, the probability of adding a feature is $1 - e^{-\epsilon} = 1 - (1 - \epsilon + O(\epsilon^2)) \approx \epsilon$, and the cost in bits of coding the fact the feature is added is roughly $-\log(\epsilon)$ bits. Different codings can be used for the feature's estimated coefficient. For example, BIC uses $\frac{1}{2}\log(n)$ bits. In section 3.4.3, we present an optimal coding of the estimated coefficients.

For simplicity assume we use 2 bits to code each feature $x$'s estimated coefficient $\hat{\beta}$ when $x$ is added to the model. The choice of 2 bits is justified in Section 4.2.2 of Chapter 4. Adding $x$ to the model reduces the model entropy by $\frac{1}{2}t^2 \log(e)$ bits where $t$ is the $t$ statistic associated with $\hat{\beta}$, as defined above. Here, and below $\log()$ is log based 2; the $\log(e)$ converts the $t^2$ to bits. Then,

---

[6]Slightly better and more complex bid selection methods such as $\epsilon_i \leftarrow w_i/(i\log(i))$ could be used, but they are statistically equivalent to the simpler one in terms of rates, and more importantly they generate tests that have no more power. We will stick with the simpler one in this chapter.

$$bits\_saved = \frac{1}{2}t^2 \log(e) - (-\log(\epsilon) + 2).$$

**Optimal Coding of Coefficients in Information-investing**

A key question is what coding scheme to use to code the coefficient of a feature which is added to the model. We describe here an "optimal" coding scheme which can be used in the information-investing criterion. The key idea is that coding an event with probability $p$ requires $\log(p)$ bits. This equivalence allows us to think in terms of distributions and thus to compute codes which handle fractions of a bit. Our goal is to find a (legitimate) coding scheme which, given a "bid" $\epsilon$, will guarantee the highest probability of adding the feature to the model. We show below that given any actual distribution $\tilde{f}_\beta$ of the coefficients, we can produce a coding corresponding to a modified distribution $f_\beta$ which produces a coding which uniformly dominates it.

Assume, for simplicity, that we increase the wealth by one bit when a feature $x_i$ with coefficient $\beta_i$ is added. Thus, when $x_i$ is added, we have

$$\log \frac{p(x_i \text{ is a beneficial feature})}{p(x_i \text{ is a spurious feature})} > 1 \text{ bit,}$$

that is, the log-likelihood decreases by more than one bit.

Let $f_{\beta_i}$ be the distribution implied by the coding scheme for $t_{\beta_i}$ if we add $x_i$ and $f_0(t_{\beta_i})$ be the normal distribution (the null model in which $x_i$ should not be added). The coding saves enough bits to justify adding a feature whenever $f_{\beta_i}(t_{\beta_i}) \geq 2f_0(t_{\beta_i})$. This happens with probability $\alpha_i \equiv p_0(\{t_{\beta_i} : f_{\beta_i}(t_{\beta_i}) \geq 2f_0(t_{\beta_i})\})$ under the null. $\alpha_i$ is the area under the tails of the null distribution.

There is no reason to have $f_{\beta_i}(t_{\beta_i}) \gg 2f_0(t_{\beta_i})$ in the tails, since this would "waste" probability or bits. Hence, the optimal coding is $f_\beta(t_{\beta_i}) = 2f_0(t_{\beta_i})$ for all the features that are likely to be added. Using all of the remaining probability mass (or equivalently, making the coding "Kraft tight") dictates the coding for the case

36

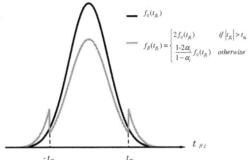

Figure 3.4: Optimal distribution $f_\beta$.

when the feature is not likely to be added. The most efficient coding to use is thus:

$$\begin{cases} f_\beta(t_{\beta_i}) = 2f_0(t_{\beta_i}) & \text{if } |t_{\beta_i}| > t_{\alpha_i} \\ f_\beta(t_{\beta_i}) = \frac{1-2\alpha_i}{1-\alpha_i} f_0(t_{\beta_i}) & \text{otherwise} \end{cases}$$

and the corresponding cost in bits is:

$$\begin{cases} \log(f_\beta(t_{\beta_i})/f_0(t_{\beta_i})) = \log(2) = 1 \text{ bit} & \text{if } |t_{\beta_i}| > t_{\alpha_i} \\ \log(f_\beta(t_{\beta_i})/f_0(t_{\beta_i})) = \log(\frac{1-2\alpha_i}{1-\alpha_i}) \approx -\alpha_i \text{ bits} & \text{otherwise.} \end{cases}$$

Figure 3.4 shows the distribution $f_\beta(t(\beta_i))$, with the probability mass transferred away from the center, where features are not added, out to the tails, where features are added.

The above equations are derived assuming that 1 bit is added to the wealth. It can be generalized to add $W_\Delta$ bits to the wealth each time a feature is added to the model. Then, when a feature is added to the model the probability of it being "beneficial" should be $2^{W_\Delta}$ times that of it being "spurious", and all of the 2's in the above equations are replaced with $2^{W_\Delta}$.

37

## 3.5 Streamwise Regression using Alpha-investing

One can define an alternate form of streamwise regression, $\alpha$-investing [46], which is phrased in terms of p-values rather than information theory. The p-value associated with a t-statistic is the probability that a coefficient of the observed size could have been estimated by chance even though the true coefficient was zero [31]. Of the three components of streamwise regression using information-investing, in $\alpha$-investing, *wealth updating* is similar, *bid selection* is identical, and *feature coding* is not required. The two different streamwise regression algorithms are asymptotically identical (the wealth update of $\alpha_\Delta - \alpha_i$ approaches the update of $W_\Delta$ as $\alpha_i$ becomes small), but differ slightly when the initial features in the stream are considered. The relation between the two methods follows from the fact that coding an event with probability $p$ requires $\log(p)$ bits. The $\alpha$-investing algorithm is shown in Figure 3.5, and the equivalence between $\alpha$-investing and information-investing is shown in Table 3.3. Wealth updating is now done in terms of $\alpha$, the probability of adding a spurious feature.

$\alpha$-investing controls the FDR bound by dynamically adjusting a threshold on the p-statistic for a new feature to enter the model [46]. Similarly to the information-investing, $\alpha$-investing adds as many features as possible subject to the FDR bound giving the minimum out-of-sample error.

The threshold, $\alpha_i$, corresponds to the probability of including a spurious feature at step $i$. It is adjusted using the wealth, $w_i$, which represents the current acceptable number of future false positives. Wealth is increased when a feature is added to the model (presumably correctly, and hence permitting more future false positives without increasing the overall FDR). Wealth is decreased when a feature is not added to the model. In order to save enough wealth to add future features, bid selection is identical to the information-investing.

More precisely, a feature is added to the model if its p-value is greater than $\alpha$. The p-value is computed by using the fact that $\Delta$log-likelihood is equivalent to a

**Input:** A vector of $y$ values (for example, labels), a stream of features $x$, $W_0$, and $\alpha_\Delta$.

{initialize}
model = {}                    //initially no features in model
$i = 1$                       // index of features
$w_1 = W_0$                   // initial prob. of false positives
**while** CPU_time_used < max_CPU_time **do**
    $x_i \leftarrow$ get_next_feature()
    $\alpha_i \leftarrow w_i/2i$
    {Is p-value of the new feature below threshold?}
    **if** $get\_p\text{-}value(x_i, model) < \alpha_i$ **then**
        model $\leftarrow$ model $\cup$ $x_i$   // add $x_i$ to the model
        $w_{i+1} \leftarrow w_i + \alpha_\Delta - \alpha_i$ // increase wealth
    **else**
        $w_{i+1} \leftarrow w_i - \alpha_i$        // reduce wealth
    **end if**
    $i \leftarrow i + 1$
**end while**

Figure 3.5: Algorithm: streamwise regression with $\alpha$-investing.

| information-investing | $\alpha$-investing |
|:---:|:---:|
| $w_i$ | $\log(w_i)$ |
| bits_saved | test statistic = $\Delta$log-likelihood |
| bits_saved > $W_\Delta$ | p-value < $\alpha_i$ |

Table 3.3: The equivalence of $\alpha$-investing and information-investing.

t-statistic. The idea of $\alpha$-investing is to adaptively control the threshold for adding features so that when new (probably predictive) features are added to the model, one "invests" $\alpha$ increasing the wealth, raising the threshold, and allowing a slightly higher future chance of incorrect inclusion of features. We increase wealth by $\alpha_\Delta - \alpha_i$. Note that when $\alpha_i$ is very small, this increase amount is roughly equivalent to $\alpha_\Delta$. Each time a feature is tested and found not to be significant, wealth is "spent", reducing the threshold so as to keep the guarantee of not adding more than a target fraction of spurious features. There are two user-adjustable parameters, $\alpha_\Delta$ and $W_0$, which can be selected to control the FDR; we set both of them to 0.5 in all of the experiments presented in this chapter.

$\alpha$-investing allows us to bound, in expectation, the relative fraction of features incorrectly and correctly added to the model. The theorem, presently below, is analogous to Theorem 1 in the information-investing case.

**Theorem 3** *Let $S_i$ be the number of beneficial features included in the model, let $V_i$ be the number of spurious features included and $w_i$ be the wealth, all at iteration $i$, and let $\alpha_\Delta < 1$ be a user selected value. Then if the algorithm in Figure 3.5 is followed:*

$$E(V_i) < (\alpha_\Delta E(S_i) + W_0)/(1 - \alpha_\Delta).$$

**Proof** The proof relies on the fact that

$$G_i \equiv (1 - \alpha_\Delta)V_i - \alpha_\Delta S_i + w_i$$

is a super-martingale. That is, $G_i$ is, in expectation, non-increasing in each iteration: $E(G_i) \leq G_{i-1}$. Thus

$$E(G_i) \leq G_1,$$

but since we start out with $V_i = 0$, and $S_i = 0$,

$$E((1 - \alpha_\Delta)V_i - \alpha_\Delta S_i + w_i) \leq W_0.$$

40

Further, since $w_i > 0$ by construction,

$$E((1 - \alpha_\Delta)V_i - \alpha_\Delta S_i) < W_0.$$

The proof that $G_i$ is a super-martingale is straightforward by considering the cases when the feature is or is not a beneficial feature and is or is not added to the estimated model. ∎

Our result, which can be written as

$$E(V_i)/(E(S_i) + W_0/\alpha_\Delta) < \alpha_\Delta/(1 - \alpha_\Delta)$$

or, as many features are added to the model,

$$E(V_i)/E(S_i) < \alpha_\Delta/(1 - \alpha_\Delta).$$

We again have an mFDR bound. Similarly to the proof presented in the information-investing case, adding as many features as possible subject to the mFDR bound gives the minimum out-of-sample error.

## 3.6 Experimental Evaluation

We compared streamwise feature selection using $\alpha$-investing against both streamwise and stepwise feature selection (see Section 3.2) using the AIC, BIC and RIC penalties on a battery of synthetic and real data sets. After a set of features are selected from the real data sets, we applied logistic regression on this feature set selected, calculated the probability of observation labels, provided a cutoff/threshold of 0.5 to classify the response labels if label values are binary and get the prediction accuracies or balance errors. (Actually, different cutoffs could be used for different loss functions.) Information-investing gives extremely similar results, so we do not report them. We used R to implement our evaluation.

### 3.6.1 Evaluation on Synthetic Data

The base synthetic data set contains 100 observations each of 1,000 features, of which 4 are predictive. We generated the features independently from a normal distribution, $N(0,1)$, with the true model being the sum of four of the features (their coefficients are one's)[7] plus noise, $N(0, 0.1^2)$. The artificially simple structure of the data (the features are uncorrelated and have relatively strong signal) allows us to easily see which feature selection methods are adding spurious features or failing to find features that should be in the model.

The results are presented in Table 3.4. As expected, AIC massively overfits, always putting in as many features as there are observations. BIC overfits severely, although less badly than AIC. RIC gives performance comparable to $\alpha$-investing. As one would also expect, if all of the beneficial features in the model occur at the beginning of the stream, $\alpha$-investing does better, giving the same error as RIC, while if all of the beneficial features in the model are last, $\alpha$-investing does (two times) worse than RIC. In practice, if one is not taking advantage of known structure of the features, one can randomize the feature order to avoid such bad performance.

Stepwise regression gave noticeably better results than streamwise regression for this problem when the penalty is AIC or BIC. Using AIC and BIC still resulted in $n$ features being added, but at least all of the beneficial features were found. Stepwise regression with RIC gave the same error of its streamwise counterpart. However, using standard code from R, the stepwise regression was *much* slower than streamwise regression. Running stepwise regression on data sets with tens of thousands of features, such as the ones presented in Table 3.5, was not possible.

One might hope that adding more spurious features to the end of a feature stream would not severely harm an algorithm's performance.[8] However, AIC and BIC, since

---

[7]Similar results are also observed, if instead of using coefficients which are strictly 0 or 1, we use coefficients that are generated in either of the two cases: (a) most coefficients are zeros and several are from Gaussian distribution; (b) all coefficients are generated from t distribution with degree of freedom of two.

[8]One would not, of course, intentionally add features known not to be predictive. However, as

| streamwise | AIC | BIC | RIC | α-invest. | α-invest. first | last |
|---|---|---|---|---|---|---|
| features | 100 | 90 | 4.3 | 4.2 | 4.6 | 3.7 |
| error | 6.13 | 1.91 | 0.33 | 0.42 | 0.33 | 0.71 |
| stepwise | AIC | BIC | RIC | | | |
| features | 100 | 100 | 4.5 | – | – | – |
| error | 0.54 | 0.54 | 0.33 | – | – | – |

Table 3.4: **AIC and BIC overfit for** $m \gg n$. The number of features selected and the out-of-sample error, averaged over 20 runs. $n = 100$ observations, $m = 1,000$ features, $m^* = 4$ beneficial features in data. Synthetic data: $x \sim N(0,1)$, $y$ is linear in $x$ with noise $\sigma^2 = 0.1$. Beneficial features are randomly distributed in the feature set except the "first" and "last" cases. "first" and "last" denote the beneficial features being first or last in the feature stream.

| m | | 1,000 | 10,000 | 100,000 | 1,000,000 |
|---|---|---|---|---|---|
| RIC | features | 4.3 | 4.0 | 4.0 | 3.4 |
| RIC | false pos. | 0.3 | 0.2 | 0.2 | 0.4 |
| RIC | error | 0.33 | 0.42 | 0.50 | 0.97 |
| α-invest. | features | 4.2 | 4.1 | 4.7 | 4.8 |
| α-invest. | false pos. | 0.3 | 0.2 | 0.7 | 0.9 |
| α-invest. | error | 0.42 | 0.42 | 0.43 | 0.45 |

Table 3.5: **Effect of adding spurious features.** The average number of features selected, false positives, and out-of-sample error (20 runs). $m^* = 4$ beneficial features, randomly distributed over the first 1,000 features. Otherwise the same model as Table 3.4.

their penalty is not a function of $m$, will add even more spurious features (if they haven't already added a feature for every observation!). RIC (Bonferroni) produces a harsher penalty as $m$ gets large, adding fewer and fewer features. As Table 3.5 and 3.6 show, α-investing is clearly the superior method in this case. Table 3.6 shows that when the number of potential features goes up to 1,000,000, RIC puts in one less beneficial feature, while streamwise regression puts the same four beneficial features plus a half of a spurious feature. Thus, streamwise regression is able to find the extra feature even when the feature is way out in the 1,000,000 features.

---

described above, there is often a natural ordering on features so that some classes of features, such as interactions, have a smaller fraction of predictive features, and can be put later in the feature stream.

| m | | 1,000 | 10,000 | 100,000 | 1,000,000 |
|---|---|---|---|---|---|
| RIC | features | 4.3 | 4.2 | 3.9 | 3.7 |
| RIC | false pos. | 0.3 | 0.3 | 0.1 | 0.6 |
| RIC | error | 0.33 | 0.42 | 0.50 | 0.97 |
| $\alpha$ invest. | features | 4.2 | 4.2 | 4.5 | 4.9 |
| $\alpha$ invest. | false pos. | 0.3 | 0.3 | 0.6 | 0.8 |
| $\alpha$ invest. | error | 0.42 | 0.42 | 0.43 | 0.42 |

Table 3.6: **Effect of adding spurious features.** The average number of features selected, false positives, and out-of-sample error (20 runs). $m^* = 4$ beneficial features: when $m = 1,000$, all four beneficial features are randomly distributed; in the other three cases, there are three beneficial features randomly distributed over the first 1,000 features and another beneficial feature randomly distributed within the feature index ranges [1001, 10000], [10001, 100000], and [100001, 1000000] when $m = 10000, 100000,$ and $1000000$ respectively. Otherwise the same model as Table 3.4 and 3.5.

## 3.6.2 Evaluation on Real Data

Tables 3.7, 3.8, and 3.9 provide a summary of the characteristics of the real data sets that we used. All are for binary classification tasks. The six data sets in Table 3.7 were taken from the UCI repository. The seven data sets in Table 3.8 are bio-medical data, in which each feature represents a gene expression value for each observation (patient with cancer or healthy donor). For example, in *aml* data set, observations consist of patients with acute myeloid leukemia and patients with acute lymphoblastic leukemia. The classification task is to identify which patient has which cancer. *ha* and *hung* are private data sets and other gene expression data sets are available to the public [32]. The NIPS data sets are from the NIPS2003 workshop [24].

The observations are shuffled and those observations which contain missing feature values are deleted. Since the gene expression data sets have large feature sets, we shuffled their original features five times (in addition to the cross validations), applied streamwise regression on each feature order, and averaged the five evaluation results. The baseline accuracy is the accuracy (on the whole data set) when predicting the majority class. The feature selection methods were tested on these

|  | cleve | internet | ionosphere | spect | wdbc | wpbc |
|---|---|---|---|---|---|---|
| features, $m$ | 13 | 1558 | 34 | 22 | 30 | 33 |
| nominal features | 7 | 1555 | 0 | 22 | 0 | 0 |
| continuous features | 6 | 3 | 34 | 0 | 30 | 33 |
| observations, $n$ | 296 | 2359 | 351 | 267 | 569 | 194 |
| baseline accuracy | 54% | 84% | 64% | 79% | 63% | 76% |

Table 3.7: **Description of the UCI data sets.**

|  | aml | ha | hung | ctumor | ocancer | pcancer | lcancer |
|---|---|---|---|---|---|---|---|
| features, $m$ | 7,129 | 19,200 | 19,200 | 2,000 | 15,154 | 12,600 | 12,533 |
| observations, $n$ | 72 | 83 | 57 | 62 | 253 | 136 | 181 |
| baseline accuracy | 65% | 71% | 63% | 65% | 64% | 57% | 92% |

Table 3.8: **Description of the gene expression data sets.** All features are continuous.

data sets using ten-fold cross-validation.

On the UCI and gene expression data sets, experiments were done on two different feature sets. The first experiments used only the original feature set. The second interleaved feature selection and generation, initially testing PCA components and the original features, and then generating interaction terms between any of the features which had been selected and any of the original features. On the NIPS data sets, since our main concern is to compare against the challenge best models, we did only the second kind of experiment.

|  | arcene | dexter | dorothea | gisette | madelon |
|---|---|---|---|---|---|
| features, $m$ | 10,000 | 20,000 | 100,000 | 5,000 | 500 |
| observations, $n$ | 100 | 300 | 800 | 6,000 | 2,000 |
| baseline accuracy | 56% | 50% | 90% | 50% | 50% |

Table 3.9: **Description of the NIPS data sets.** All features are nominal.

45

On UCI data sets (Table 3.10, Table 3.11, Figure 3.6)[9] , when only the original feature set is used, paired two-sample t-tests show that $\alpha$-investing has better performance than streamwise AIC and BIC only on two of the six UCI data sets: the *internet* and *wpbc* data sets. On the other data sets, which have relatively few features, the less stringent penalties do as well as or better than streamwise regression. When interaction terms and PCA components are included, $\alpha$-investing gives better performance than streamwise AIC on five data sets, than streamwise BIC on three data sets, and than streamwise RIC on two data sets. In general, when the feature set size is small, there is no significant difference in the prediction accuracies between $\alpha$-investing and the other penalties. When the feature set size is larger (that is, when new features are generated) $\alpha$-investing begins to show its superiority over the other penalties.

On the UCI data sets (Table 3.10, Table 3.11, Figure 3.6), we also compared streamwise regression with $\alpha$-investing[10] with stepwise regression. Paired two-sample t-tests show that when the original feature set is used, $\alpha$-investing does not differ significantly from stepwise regression. $\alpha$-investing has better performance than stepwise regression in 5 cases, and worse performance in 3 cases. (Here a "case" is defined as a comparison of $\alpha$-investing and stepwise regression under a penalty, that is, AIC or BIC or RIC, on a data set.) However, when interaction terms and PCA components are included, $\alpha$-investing gives better performance than stepwise regression in 9 cases, and worse performance in none of the cases. Thus, in our tests, $\alpha$-investing

---

[9]In Figure 3.6, a small training set size of 50 was selected to make sure the problems were difficult enough that the methods gave clearly different results. The right columns graphs differs from the left ones in that: (1) we generated PCA components from the original data sets and put them at the front of the feature sets; (2) after the PCA component "block" and the original feature "block", there is an interaction term "block" in which the interaction terms are generated using the features selected from the first two feature blocks. This kind of feature stream was also used in the experiments on the other data sets. We were unable to compute the stepwise regression results on the internet data set using the software at hand when interaction terms and PCA components were included giving millions of potential features with thousands of observations. It is indicative of the difficulty of running stepwise regression on large data sets.

[10]In later text of this section, for simplicity, we use "$\alpha$-investing" to mean the "streamwise regression with $\alpha$-investing".

is comparable to stepwise regression on the smaller data sets and superior on the larger ones.

On the UCI data sets (Table 3.12), $\alpha$-investing was also compared with support vector machines (SVM), neural networks (NNET), and decision tree models (TREE). In all cases, standard packages available with R were used[11]. No doubt these could be improved by fine tuning parameters and kernel functions, but we were interested in seeing how well "out-of-the-box" methods could do. We did not tune any parameters in streamwise regression to particular problems either. Paired two-sample t-tests show that $\alpha$-investing has better performance than NNET on 3 out of 6 data sets, and than SVM and TREE on 2 out of 6 data sets. On the other data sets, streamwise regression doesn't have significant better or worse performance than NNET, SVM, or TREE. These tests shows that the performance of streamwise regression is at least comparable to those of SVM, NNET, and TREE.

On the gene expression data sets (Table 3.13, Table 3.14, Figure 3.7), when comparing $\alpha$-investing with streamwise AIC, streamwise BIC, and streamwise RIC, paired two-sample t-tests show that when the original features are used, the performances of $\alpha$-investing and streamwise RIC don't have significant difference on any of the data sets. But when interaction terms and PCA components are included , RIC is often too conservative to select even only one feature, whereas $\alpha$-investing has stable performance and the t-tests show that $\alpha$-investing has significant better prediction accuracies than streamwise RIC on 5 out of 7 data sets. Note that, regardless of whether or not interaction terms and PCA components are included, $\alpha$-investing always has much higher accuracy than streamwise AIC and BIC.

The standard errors (SE) of prediction accuracies in shuffles gave us sense of the approach sensitivity to the feature order. When the original features are used, $\alpha$-investing has a maximum SE of four percent on *pcancer* and its other SEs are less

---

[11]Please find details at http://cran.r-project.org/doc/packages for SVM (e1071.pdf), NNET (VR.pdf), and TREE (tree.pdf). SVM uses the radial kernel. NNET uses feed-forward neural networks with a single hidden layer. TREE grows a tree by binary recursive partitioning using the response in the specified formula and choosing splits from the terms of the right-hand-side.

than two percents in accuracy. When PCA components and interaction terms are included, $\alpha$-investing has a maximum SE of two percents on *ha* and its other SEs are around or less than one percent. Streamwise RIC has similar SEs as $\alpha$-investing has, but streamwise AIC and BIC usually have one percent higher SEs than $\alpha$-investing and streamwise RIC. We can see that the feature shuffles don't change the performance much on most of gene expression data sets.

When PCA components and interaction terms are included and the original feature set is sorted in advance by feature value variance (one simple way of making use of the ordering in the stream), the prediction accuracy of $\alpha$-investing on *hung* is increased from 79.3% to 86.7%; for the other gene expression data, sorting gave no significant change.

Also note that, for streamwise AIC, BIC, and RIC, adding interaction terms and PCA components often hurts. In contrast, the additional features have not much effect on $\alpha$-investing. With these additional features, the prediction accuracies of $\alpha$-investing are improved or kept the same on 4 out of 6 UCI data sets and 5 out of 7 gene data sets.

On the gene expression data sets (Table 3.13, Table 3.14, Figure 3.7), we also compared $\alpha$-investing with stepwise regression. The results show that, $\alpha$-investing is competitive with stepwise regression with the RIC penalty. Stepwise regression with AIC or BIC penalties gives inferior performance.

On the NIPS data sets (Table 3.15), we compared $\alpha$-investing against results reported on the NIPS03 competition data set using other feature selection methods [25]. Table 3.15 shows the results we obtained, and compares them against the two methods which did best in the competition. These methods are *BayesNN-DFT* [33, 34], which combines Bayesian neural networks and Bayesian clustering with a Dirichlet diffusion tree model and *greatest-hits-one* [23], which normalizes the data set, selects features using distance information, and classifies them using a perceptron or SVM.

Different feature selection approaches such as those used in *BayesNN-DFT* can be contrasted based on their different levels of greediness. *Screening* methods or filters look at the relation between $y$ and each $x_i$ *independently*. In a typical screen, one computes how predictive each $x_i$ $(i = 1...m)$ is of y (or how are they correlated), or the mutual information between them, and all features above a threshold are selected. In an extension of the simple screen, FBFS [13] looks at the mutual information $I(y; x_i | x_j)$ $(i, j = 1...m)$, that is, the effect of adding a second feature after one has been added.[12] Streamwise and stepwise feature selection are one step less greedy, sequentially adding features by computing $I(y; x_i | Model)$.

*BayesNN-DFT* uses a screening method to select features, followed by a sophisticated Bayesian modeling method. Features were selected using the union of three univariate significance test-based screens [35]: correlation of class with the ranks of feature values, correlation of class with a binary form of the feature (zero/nonzero), and a runs test on the class labels reordered by increasing feature value. The threshold was selected by comparing each to the distribution found when permuting the class labels. This richer feature set of transformed variables could, of course, be used within the streamwise feature selection setting, or streamwise regression could be used to find an initial set of features to be provided to the Bayesian model.

*greatest-hits-one* applied margin based feature selection on data sets *arcene* and *madelon*, and used a simple infogain ranker to select features on data sets *dexter* and *dorothea*. Assuming a fixed feature set size, a generalization error bound is proved for the margin based feature selection method [23].

Table 3.15 shows that we mostly get comparable accuracy to the best-performing of the NIPS03 competition methods, while using a small fraction of the features. Many of the NIPS03 contestants got far worse performance, without finding small feature sets [36]. When SVM is used on the features selected by streamwise regression, the errors on *arcene*, *gisette*, and *madelon* are reduced further to 0.151, 0.021,

---

[12]FBFS has been developed only for binary features, but could be easily extended.

|  | cleve | internet | ionosphere |
|---|---|---|---|
| step. (AIC) | 75.9±1.1 (6.3) | 91.0±0.5 (2.9) | 79.8±0.9 (7.7) |
| step. (BIC) | 76.6±1.0 (4.9) | 91.0±0.5 (2.9) | 82.7±1.4 (3.0) |
| step. (RIC) | 73.7±1.7 (3.4) | 88.2±1.0 (0.8) | 83.6±1.0 (2.1) |
| stream. (AIC) | 77.9±0.9 (6.1) | 86.4±0.7 (14.7) | 83.2±0.9 (7.9) |
| stream. (BIC) | 77.3±1.1 (4.8) | 86.8±0.9 (10.2) | 80.1±2.5 (4.0) |
| stream. (RIC) | 76.0±1.0 (4.2) | 90.3±0.2 (3.6) | 78.3±2.8 (1.6) |
| stream. ($\alpha$-invest.) | 72.2±2.1 (2.7) | 90.1±0.1 (3.7) | 83.4±0.9 (2.4) |

|  | spect | wdbc | wpbc |
|---|---|---|---|
| step. (AIC) | 76.2±1.3 (5.0) | 90.2±0.7 (4.1) | 68.7±2.2 (4.4) |
| step. (BIC) | 71.2±1.6 (3.1) | 90.2±0.7 (4.1) | 69.9±1.6 (2.2) |
| step. (RIC) | 69.6±1.8 (2.5) | 91.3±0.6 (3.1) | 72.2±0.9 (1.2) |
| stream. (AIC) | 75.2±1.0 (6.9) | 91.5±1.4 (9.6) | 69.9±1.9 (4.0) |
| stream. (BIC) | 75.6±0.8 (3.5) | 92.4±0.8 (5.5) | 71.8±1.5 (2.1) |
| stream. (RIC) | 75.3±1.6 (2.2) | 92.4±0.7 (3.4) | 74.0±0.8 (1.0) |
| stream. ($\alpha$-invest.) | 74.0±1.6 (2.1) | 91.6±0.5 (3.4) | 75.1±0.5 (0.9) |

Table 3.10: **UCI data validation accuracy on raw features for different penalties.** Training size is 50. A small training set size was selected to make sure the problems were difficult enough that the methods gave clearly different results. The number before ± is the average accuracy on 10 cross-validations; the number immediately after ± is the standard deviation of the average accuracy. The number in parentheses is the average number of features selected by SFS or stepwise regression.

0.214 respectively. One could also apply more sophisticated methods, such as the Bayesian models which *BayesNN-DFT* used, to our selected features.

There is only one data set, *madelon*, where streamwise regression gives substantially higher error than the other methods. This may be partly due to the fact that madelon has substantially more observations than features, thus making streamwise regression (when not augmented with sophisticated feature generators) less competitive with more complex models. Madelon is also the only synthetic data set in the NIPS03 collection, and so its structure may benefit more from the richer models than typical real data.

|  | cleve | internet | ionosphere |
|---|---|---|---|
| step. (AIC) | 73.9±1.3 (6.7) | NA | 78.7±1.2 (8.1) |
| step. (BIC) | 76.3±0.9 (5.8) | NA | 79.5±2.3 (5.2) |
| step. (RIC) | 70.7±1.7 (2.2) | NA | 79.8±1.7 (1.1) |
| stream. (AIC) | 70.1±1.3 (27.5) | 82.7±1.7 (37.7) | 61.7±3.2 (44) |
| stream. (BIC) | 75.4±0.9 (10.3) | 85.5±1.6 (21.3) | 83.5±0.9 (8.4) |
| stream. (RIC) | 70.2±1.7 (2.5) | 90.5±0.1 (1.5) | 74.6±2.5 (0.8) |
| stream. ($\alpha$-invest.) | 74.1±1.2 (4.2) | 89.6±0.4 (9.2) | 85.2±0.5 (3.3) |
|  | spect | wdbc | wpbc |
| step. (AIC) | 78.9±1.2 (3.3) | 89.9±0.6 (4.2) | 64.7±2.7 (5.9) |
| step. (BIC) | 78.7±1.5 (2.0) | 90.9±0.9 (3.6) | 65.4±2.3 (3.7) |
| step. (RIC) | 81.4±0.3 (1.0) | 90.2±0.7 (1.7) | 73.6±1.0 (0.4) |
| stream. (AIC) | 78.5±1.6 (6.4) | 69.2±3.3 (44.6) | 66.6±2.3 (18.9) |
| stream. (BIC) | 80.9±0.4 (2.4) | 91.1±1.0 (7.8) | 71.1±1.6 (3.9) |
| stream. (RIC) | 81.6±0.3 (1.8) | 93.3±0.6 (1.9) | 74.3±0.9 (0.4) |
| stream. ($\alpha$-invest.) | 81.1±0.4 (1.7) | 94.4±0.5 (3.5) | 73.7±0.9 (0.8) |

Table 3.11: **UCI data validation accuracy on PCAs, raw features, and interaction terms for different penalties.** Training size is 50. Same format as earlier tables. This table differs from Table 3.10 in that: (1) we generated PCA components from the original data sets and put them at the front of the feature sets; (2) after the PCA feature "block" and the original feature "block", there is an interaction term "block" in which the interaction terms are generated using the features selected from the first two feature blocks. "NA" means that we were unable to compute the results using the software at hand. See Section 3.3.

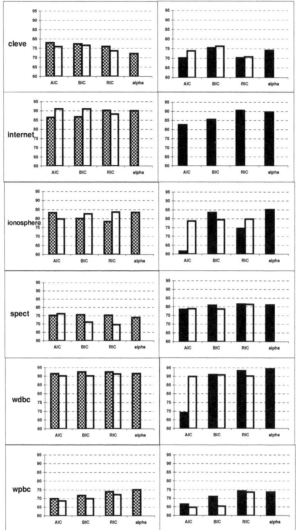

Figure 3.6: **UCI Data. Streamwise vs. stepwise validation accuracy for different penalties.** Training size is 50. The average accuracy is on 10 cross-validations. The black-dot and solid black bars are the average accuracies using streamwise regressions. The transparent bars are the average accuracies using stepwise regressions. Raw features are used in the left column graphs. Additional interaction terms and PCA components are used in the right column graphs. Please see Footnote 9 for additional information about this figure. Section 3.6.2 gives the results of paired two-sample t-tests.

|  | cleve | internet | ionosphere |
|---|---|---|---|
| stream. ($\alpha$-invest.) | 84.3±1.8 (8.5) | 96.5±0.3 (166) | 91.4±1.8 (23) |
| SVM | 82.0±2.0 | 93.4±0.6 | 92.2±1.7 |
| NNET | 70.3±4.5 | 84.2±0.9 | 91.7±1.9 |
| TREE | 76.0±3.3 | 96.5±0.5 | 86.7±1.8 |

|  | spect | wdbc | wpbc |
|---|---|---|---|
| stream. ($\alpha$-invest.) | 82.2±2 (2) | 95.1±0.6 (37) | 77±3.4 (4.4) |
| SVM | 81.5±2.4 | 96.3±0.8 | 76.5±4.9 |
| NNET | 78.9±2.2 | 68.8±4.5 | 75.0±5.0 |
| TREE | 81.1±1.9 | 94.2±0.9 | 74.0±3.0 |

Table 3.12: **Comparison of streamwise regression and other methods on UCI data.** Average accuracy using 10-fold cross validation. The number before ± is the average accuracy; the number immediately after ± is the standard deviation of the average accuracy. The number in parentheses is the average number of features used by the streamwise regression, and these features includes PCA components, raw features, and interaction terms (see Footnote 9 for the details of this kind of feature stream). SVM, NNET, and TREE use the whole raw feature set. Section 3.6.2 gives the results of paired two-sample t-tests.

|  | aml | ha | hung |
|---|---|---|---|
| step. (AIC) | 95.0±2.8 | 87.8±3.1 | 71.7±7.0 |
| step. (BIC) | 95.0±2.8 | 87.8±3.1 | 71.7±7.0 |
| step. (RIC) | 93.8±2.8 | 88.9±2.3 | 75.0±4.5 |
| stream. (AIC) | 65.3±3.8 | 59.3±2.5 | 60.0±4.9 |
| stream. (BIC) | 71.5±1.0 | 71.1±2.2 | 69.0±4.1 |
| stream. (RIC) | 91.2±1.0 | 79.1±0.9 | 80.3±2.4 |
| stream. ($\alpha$-invest.) | 90.8±1.9 | 77.8±1.3 | 81.0±1.8 |

|  | ctumor | ocancer | pcancer | lcancer |
|---|---|---|---|---|
| step. (AIC) | 64.3±5.3 | 100.0±0.0 | 72.9±3.2 | 97.4±1.2 |
| step. (BIC) | 64.3±5.3 | 100.0±0.0 | 73.6±3.0 | 97.4±1.2 |
| step. (RIC) | 81.4±6.0 | 100.0±0.0 | 72.9±4.1 | 98.4±0.8 |
| stream. (AIC) | 62.6±2.9 | 79.4±4.1 | 57.7±1.1 | 81.1±1.4 |
| stream. (BIC) | 63.7±2.5 | 95.8±1.7 | 64.4±2.9 | 93.6±1.2 |
| stream. (RIC) | 80.0±1.0 | 100.0±0.0 | 66.9±3.0 | 98.5±0.3 |
| stream. ($\alpha$-invest.) | 79.1±1.2 | 100.0±0.0 | 65.6±4.0 | 98.3±0.2 |

Table 3.13: **Gene expression data validation accuracy on raw features for different penalties.** Standard forward 10-fold cross validations. Same format as earlier tables.

|  | aml | ha | hung |
|---|---|---|---|
| stream. (AIC) | 82.2±1.2 | 60±3.3 | 64.7±2.3 |
| stream. (BIC) | 85.5±1.6 | 69.3±1.4 | 72±1.9 |
| stream. (RIC) | 92.5±0 | 77.1±0.8 | 63.3±0 |
| stream. (α-invest.) | 96.2±0.4 | 82.7±2.4 | 79.3±1.2 |

|  | ctumor | ocancer | pcancer | lcancer |
|---|---|---|---|---|
| stream. (AIC) | 58.6±2.3 | 84.9±1.9 | 60.3±0.8 | 85.9±0.6 |
| stream. (BIC) | 58.3±3.8 | 99.4±0.3 | 65.4±1.6 | 96±0.4 |
| stream. (RIC) | 64±0.5 | 100±0 | 56.4±0 | 97.8±0.4 |
| stream. (α-invest.) | 73.4±1 | 100.0±0.0 | 80.1±0.7 | 98.2±0.3 |

Table 3.14: **Gene expression data validation accuracy on PCAs, raw features, and interaction terms for different penalties.** Standard forward 10-fold cross validations. Same format as earlier tables. The feature generation is same as earlier tables.

|  | arcene | dexter | dorothea | gisette | madelon |
|---|---|---|---|---|---|
| α-invest. error | 0.176 | 0.067 | 0.090 | 0.037 | 0.295 |
| α-invest. features | 8 | 21 | 8 | 72 | 24 |
| greatest-hits-one error | 0.172 | 0.053 | 0.109 | 0.030 | 0.086 |
| greatest-hits-one features | 10,000 | 1,400 | 300 | 5,000 | 18 |
| BayesNN-DFT error | 0.133 | 0.039 | 0.085 | 0.013 | 0.072 |
| BayesNN-DFT features | 10,000 | 303 | 100,000 | 5,000 | 500 |

Table 3.15: **Comparison of streamwise regression and other methods on NIPS data.** *error* is the "balanced error"[24]; *features* is the number of features selected by models.

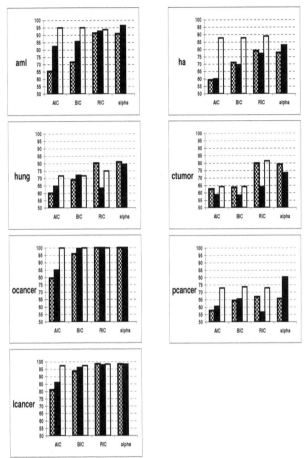

Figure 3.7: **Gene expression data. Streamwise vs. stepwise validation accuracy for different penalties.** Average accuracy using 10-fold cross validation. The black-dot bars are the average accuracies using streamwise regressions on raw features. The solid black bars are the average accuracy using streamwise regressions with PCA components, raw features, and interaction terms (see Footnote 9 for the details of this kind of feature stream). The transparent bars are the average accuracies using stepwise regressions on raw features. Section 3.6.2 gives the results of paired two-sample t-tests.

## 3.7 Discussion: Statistical Feature Selection

Recent developments in statistical feature selection take into account the size of the feature space, but only allow for finite, fixed feature spaces, and do not support streamwise feature selection. The risk inflation criterion (RIC) produces a model that possesses a type of competitive predictive optimality. RIC chooses a set of features from the potential feature pool so that the loss of the resulting model is within a factor of $\log(m)$ of the loss of the best such model. In essence, RIC behaves like a Bonferroni rule [14]. Each time a feature is considered, there is a chance that it will enter the model even if it is merely noise. In other words, the tested null hypothesis is that the proposed feature does not improve the prediction of the model. Doing a formal test generates a p-value for this null hypothesis. Suppose we only add this feature if its p-value is less than $\alpha_j$ when we consider the $j$th feature. Then the Bonferroni rule keeps the chance of adding even one extraneous feature to less than, say, 0.05 by constraining $\sum \alpha_j \leq 0.05$.

Bonferroni methods like RIC are conservative, limiting the ability of a model to add features that improve its predictive accuracy. The connection of RIC to $\alpha$-spending rules leads to a more powerful alternative. An $\alpha$-spending rule is a multiple comparison procedure that bounds its cumulative type one error rate at a small level, say 5%. For example, suppose one has to test the $m$ hypotheses $H_1, H_2, \ldots, H_m$. If we test the first using level $\alpha_1$, the second using level $\alpha_2$ and so forth with $\sum_j \alpha_j = 0.05$, then we have only a 5% chance of falsely rejecting one of the $m$ hypotheses. If we associate each hypothesis with the claim that a feature improves the predictive power of a regression, then we are back in the situation of a Bonferroni rule for feature selection. Bonferroni methods and RIC simply fix $\alpha_j = \alpha/m$ for each test.

Alternative multiple comparison procedures control a different property. Rather than controlling the cumulative $\alpha$ (also known as the family wide error rate), they control the so-called false discovery rate [4]. Control of the FDR at 5% implies that at most 5% of the rejected hypotheses are false positives. In feature selection, this implies that of the included features, at most 5% degrade the accuracy of the model. The Benjamini-Hochberg method for controlling the FDR suggests the $\alpha$-investing rule used in streamwise regression, which keeps the FDR below $\alpha$: order the p-values of the independents tests of $H_1, H_2, \ldots, H_m$ so that $p_1 \leq p_2 \leq \cdots p_m$. Now find the largest p-value for which $p_k \leq \alpha/(m - k)$ and reject all $H_i$ for $i \leq k$. Thus, if the smallest p-value $p_1 \leq \alpha/m$, it is rejected. Rather than compare the second largest p-value to the RIC/Bonferroni threshold $\alpha/m$, reject $H_2$ if $p_2 \leq 2\alpha/m$. Our proposed $\alpha$-investing rule adapts this approach to evaluate an infinite sequence of features. There have been many papers that looked at procedures of this sort for use in feature selection from an FDR perspective [1], an empirical Bayesian perspective [22, 29], an information theoretical perspective [16], or simply a data mining perspective [17]. But all of these require knowing the entire list of possible features ahead of time. Further, most of them assume that the features are orthogonal and hence tacitly assume that $m < n$. Obviously, the Benjamini-Hochberg method fails as $m$ gets large; it is a batch-oriented procedure.

The $\alpha$-investing rule of streamwise regression controls a similar characteristic. Framed as a multiple comparison procedure, the $\alpha$-investing rule implies that, with high probability, no more than $\alpha$ times the number of rejected tests are false positives. That is, the procedure controls a difference rather than a rate. As a streamwise feature selector, if one has added, say, 20 features to the model, then with high probability (tending to 1 as the number of accepted features grows) no more than 5% (that is, one feature in the case of 20 features) are false positives.


## 3.8   Summary

A variety of machine learning algorithms have been developed over the years for online learning where *observations* are sequentially added. Algorithms such as the streamwise regression presented in this chapter, which are online in the *features* being used are much less common. For some problems, all features are known in advance, and a large fraction of them are predictive. In such cases, regularization or smoothing methods work well and streamwise feature selection does not make sense. For other problems, selecting a small number of features gives a much stronger model than trying to smooth across all potential features. (See [28] and [24] for a range of feature selection problems and approaches.) For example, in predicting what journal an article will be published in, we find that roughly 10−20 of the 80,000 features we examine are selected [37]. For the problems in citation prediction and bankruptcy prediction that we have looked at, generating potential features (for example, by querying a database or by computing transformations or combinations of the raw features) takes far more time than the streamwise feature selection. Thus, the flexibility that streamwise regression provides to dynamically decide which features to generate and add to the feature stream provides potentially large savings in computation.

Empirical tests show that for the smaller UCI data sets where stepwise regression can be done, streamwise regression gives comparable results to stepwise regression or techniques such as decision trees, neural networks, or SVMs. However, unlike stepwise regression, streamwise regression scales well to large feature sets, and unlike the AIC, BIC and RIC penalties or simpler variable screening methods which use univariate tests, streamwise regression with information-investing or $\alpha$-investing works well for all values of number of observations and number of potential features. Key to this guarantee is controlling the FDR by adjusting the threshold on the information gain or p-value necessary for adding a feature to the model. Fortunately, given any software which incrementally considers features for addition and calculates their p-value or entropy reduction, streamwise regression using information-investing or $\alpha$-investing is extremely easy to implement. For linear and logistic regression, we have found that streamwise regression can easily handle millions of features.

58

The results presented here show that streamwise feature selection is highly competitive even when there is no prior knowledge about the structure of the feature space. Our expectation is that in real problems where we do know more about the different kinds of features that can be generated, streamwise regression will provide even greater benefit.

# Chapter 4

# Multiple Inclusion Criterion (MIC) for Feature Selection in Simultaneous Regressions

## 4.1 Introduction

One often wants to use estimation techniques which "borrow strength" between different regressions, in the belief that they are likely to share common sets of features or substructures [7, 19, 45, 8, 3]. When one is trying to predict a set of related responses, be they different weather measurements, different measures of the cost of health care for a person, or gene expression levels under different conditions, it is likely that methods which borrow strength between the different predictive regressions will give better performance than independent predictions.

Consider, for example, predicting gene expressions in different tissues as a function of the motifs in the promoter sequence of each gene. Hence, each observation is a gene, with a response vector of expression levels in different tissues. Each gene has associated with it a large set of features, which consist of the counts of each sequence of the four letter nucleotide alphabet of length $5 \sim 15$.

Some methods such as neural networks automatically share parameters in a model that is partially common between the different responses. Other methods such as Curds and Whey [7] or PLS [45, 20] explicitly build shared parameter models. Similar issues and approaches arise in multi-task learning [8, 3], where the goal is to use a set of related responses to improve the prediction accuracy of a single target response, often using neural networks.

This chapter looks at how to make use of multiple responses in a slightly different context, where there are a large number of spurious (non-predictive) features, and thus feature selection is likely to be of major importance.

There is a long tradition of feature selection in the single response setting, including penalized likelihood methods [5] which are preferred to cross-validation by many statisticians, as they do not require multiple re-trainings of the model and they have attractive theoretical properties. Penalized likelihood can be represented as:

$$\text{score} = -2\log\left(\text{likelihood}\right) + F \times q \tag{4.1}$$

where $F$ is a function designed to penalize model complexity, and $q$ represents the number of features currently included in the model at a given point. Table 4.1 contains the definitions which we use throughout this chapter. The first term in the equation represents a measure of the in-sample (training) error given the model, while the second is a model complexity penalty.

MIC is a penalized likelihood method based on the principle of Minimum Description Length (MDL) [40]. In MDL, both sender and receiver are assumed to know the feature matrix $X$'s, and the sender wants to construct models using $X$ to describe the response matrix $Y$ and send a message concerning the models to the receiver so that the receiver can reconstruct $Y$ using the models. We call the length of this message in bits as the *description length*. In this chapter, we use the term *total description length* (TDL) to indicate that the length of message is for the $h$ regression models.

In MIC, the two terms in equation 4.1 are two parts of the TDL. Let $S_E$ be the

61

| Symbol | Meaning |
|--------|---------|
| $n$ | Number of observations |
| $m$ | Number of candidate features |
| $m^*$ | Number of beneficial features in the candidate feature set |
| $q$ | Number of features currently included in a regression |
| $h$ | Number of simultaneous regressions |
| $k$ | Number of regressions into which a feature is added |
| $i$ | Index of regression model |
| $j$ | Index of feature |
| $v$ | Index of observation |

Table 4.1: Symbols used throughout this chapter and their definitions.

number of bits for encoding the residual errors given the models, and let $S_M$ be the number of bits for encoding the models. Therefore, we have the TDL:

$$S = S_E + S_M. \qquad (4.2)$$

In the setting of multiple responses[1], we select features for $h$ regressions simultaneously, and the $h$ models which have the minimum TDL, $S$, are selected. When we evaluate a feature, we want to maximize the reduction of TDL incurred by adding this feature to a subset $k$ regressions $(1 \leq k \leq h)$

$$\Delta S^k = \Delta S_E^k - \Delta S_M^k \qquad (4.3)$$

where $\Delta S_E^k > 0$, and $\Delta S_M^k > 0$. $\Delta S_E^k$ is the number of saved bits in describing residual error due to data likelihood increase given the new feature, and $\Delta S_M^k$ is the extra bits for encoding the model, i.e., bits for encoding the new feature.[2]

The penalty $F$ (Equation 4.1) of penalized likelihood methods include AIC [2], BIC [41], and RIC [9, 14]. In MIC, AIC, BIC, and RIC are components of $\Delta S_M^k$ in equation 4.3, and they correspond to an extra cost of 2, $\log(n)$ and $2\log(m)$

---

[1]We use "multiple responses" instead of "multiple tasks" to emphasize the difference from other multiple tasks problems in which each task has a different feature matrix; in our setting, all tasks share one feature matrix.

[2]$\Delta S_E^k$ is always greater than zero, because even a spurious feature will increase the data likelihood more or less.

bits per feature added, respectively; AIC and BIC represents the bits for encoding the regression coefficients and RIC represents the bits for encoding the index of the features that are included (see Section 4.2 for detail).

In Section 4.2.3, we compare three cases in stepwise regression: 1) a feature is added to all regressions or none; 2) features are added independently to each regression; 3) features can be added to a subset of the regressions, but the regressions share strength. We first describe this latter scheme in detail.

When a feature $x_j$ is evaluated, we try to add this feature to each of the $h$ regression models, giving a new set of residuals. Then we decide how many (value of $k$) and which regressions should actually add the feature by computing the reduction of TDL:

$$\Delta S_j^k = \Delta S_{jE}^k - \Delta S_{jM}^k \tag{4.4}$$

where $j$ is the index of current feature. At each step of the stepwise regression, all possible (remaining) features are evaluated, their corresponding maximum $\Delta S_j^k$ are calculated. If no $\Delta S_j^k$ is greater than a specified threshold, all $h$ stepwise regressions stop; if there are multiple $\Delta S_j^k$ greater than the threshold, the maximum $\Delta S_j^k$ is selected and the $j$th feature is added into the corresponding $k$ regression models.

Let us describe how to select $k$ models for the $j$th feature which is under evaluation, i.e., how to find a maximum $\Delta S_j^k$ for a fixed $j$. For each value of $k$, there are $\binom{h}{k}$ combinations of regressions. If we evaluate all $\binom{h}{k}$ combinations and select the one which gives the maximum $\Delta S_j^k$, the computation would be unaffordable when $m$ and $h$ are large. In our implementation, we use a stepwise-like selection procedure: let $k = 1$, there are $h$ candidates (regressions) in candidate set from which the best one giving maximum $\Delta S_j^k$ is selected; let $k = 2$, each candidate of candidate set is a combination of last step's selection and one of remaining $h - 1$ regressions, and from the candidate set the best one is selected; let $k = 3$, each element of candidate set is a combination of last step's selection and one of remaining $(h - 2)$ regressions, and from the candidates set the best one is selected; so on.

## 4.2 MIC Coding

### 4.2.1 Code $\Delta S_{jE}^k$

If the regressions are independent, $\Delta S_{jE}^k$ is sum of the changes in negative loglike-lihood of the $h$ regressions, i.e., $\Delta S_{jE}^k = -\sum_{i=1}^{h} \log L(y_i|M_i)$, where $M_i$ is the $i$th regression model for response $y_i$. We assume that the same feature may show up in a subset of regressions, hence the residuals at each iteration are correlated with each other; to calculate $\Delta S_{jE}^k$, we take into account the covariance matrix given by the residual matrix.

Let $\mathbf{E}$ be the residual error matrix defined as

$$\mathbf{E} = \mathbf{Y} - \hat{\mathbf{Y}} \tag{4.5}$$

where $\mathbf{Y}$ and $\hat{\mathbf{Y}}$ are the response and prediction $n \times h$ matrices respectively, let $\boldsymbol{\Sigma}_{h \times h}$ be the covariance matrix of $\mathbf{E}$, and let $\epsilon_v$ ($v = 1, 2, ..., n$) be the $v$th row of $\mathbf{E}$ and $\epsilon_v \sim N(0, \boldsymbol{\Sigma}_{h \times h})$. We have

$$P(\epsilon_v) = \frac{1}{\sqrt{2\pi\,|\boldsymbol{\Sigma}_{h \times h}|}} e^{-\frac{\epsilon_v^T \boldsymbol{\Sigma}_{h \times h}^{-1} \epsilon_v}{2}}, \tag{4.6}$$

in which $(.)^T$, $(.)^{-1}$, and $|.|$ are the matrix transpose, inverse, and determinant, respectively. Thus

$$S_E = -\log \prod_{v=1}^{n} P(\epsilon_v) = n \log \sqrt{2\pi\,|\boldsymbol{\Sigma}_{h \times h}|} + \frac{1}{2} \sum_{v=1}^{n} \epsilon_v^T \boldsymbol{\Sigma}_{h \times h}^{-1} \epsilon_v. \tag{4.7}$$

Note that the superscript $k$ in $\Delta S_{jE}^k$ means the reduction is incurred by adding a new feature to the $k$ regressions, but the calculation of $\Delta S_{jE}^k$ is over all $h$ regressions, i.e., the whole residual error $\mathbf{E}$ is taken into account.

### 4.2.2 Code $\Delta S_{jM}^k$

To describe $\Delta S_{jM}^k$ when a feature is added, MIC uses a three-part coding scheme: Let $\ell_I$ be code length for encoding the fact that the feature will be included in a subset

$k$ of models, let $\ell_H$ be the code length for specifying how many and which models include the feature, and let $\ell_\Theta$ be the code length for encoding the $k$ coefficients. This yields:

$$\Delta S_{jM}^k = \ell_I + \ell_H + \ell_\Theta. \tag{4.8}$$

The implementation in this book uses uniform distributions in specifying $\ell_I$ and $\ell_H$, which gives the code a simple form and works well in practice. We now consider different possible coding schemes for $\ell_I$, $\ell_H$, and $\ell_\Theta$.

**Code $\ell_I$**

For most of data or feature sets, little is known about how the locations of true features are distributed, we can assume a uniform distribution, i.e., if a feature $x_j$ is a beneficial feature, its index $j$ is uniformly distributed in the finite collection of integers $1, 2, ..., m$, where $m$ is the number of candidate features. Under this assumption, the MIC code extends RIC to the multiple response setting as follows:

$$\ell_I = \log(m) \tag{4.9}$$

bits are used to encode feature index.

RIC often uses no bits to code the coefficients of the features that are added. This is based on the assumption that $m$ is so large that the $\log(m)$ term dominates and the cost of specifying coefficients can be ignored. This assumption is not valid in the multiple response setting. In the MIC context, the number of regression models $h$ could be large, and if a feature is added to $k$ $(1 \le k \le h)$ regressions, the cost of encoding the $k$ coefficients may be a major part of the cost (see Section 4.2.2).

**Code $\ell_H$**

Another question is then how to code how many and which regressions will include the feature, i.e., what is $\ell_H$ equal to. We first code how many regressions will include the feature. This can be done using $\log(h)$ bits, which assumes that $k$, the number of regressions which add the feature, is from a uniform discrete distribution on $[1, h]$. In our implementation, we use this method and assumption. Alternatively, we could use $2 \log(k)$ bits to code it, which assumes that $k$ is distributed such that $P(k) = \frac{1}{k^2+1}$.

We then need to code which regressions will add the feature. A uniform distribution is also assumed, i.e, any combination from a total $\binom{h}{k}$ of combinations has the same probability of being the optimal one. Under this assumption, we use $\log \binom{h}{k}$ bits, where $\binom{h}{k}$ is the number of combinations by choosing $k$ from $h$.

Therefore, in our implementation,

$$\ell_H = \log(h) + \log \binom{h}{k}. \tag{4.10}$$

**Code $\ell_\Theta$: BIC-style ("spike-and-slab" prior) and AIC-style (universal prior)**

The coefficients of a model are estimated by maximizing the likelihood implied by the model. The question is how to code a coefficient, i.e., the maximum likelihood estimator $\theta$. We use 2 bits for each coefficient, which is similar to the AIC criterion. The choice of 2 bits is justified as follow.

[39] proposed to approximate $\theta$ using a grid resolved to the nearest standard error. That is, before coding $\theta$, we need to round it to the nearest coordinate of the form $\theta_0 + \tilde{z}SE(\theta)$, where $\theta_0$ is the default value in our null hypothesis, $SE(\theta)$ is the standard error of $\theta$, and $\tilde{z}$ is a rounded integer value of $z$ score. The coefficient $\theta$ is located by the rounded $z$-score on a grid centered at the default value $\theta_0$. The number of bits for coding $\tilde{z}$ is what we need to code $\theta$. This approximation gives negligible loss in data compression. We describe below two prior distribution of $\tilde{z}$ [42, 15].

In the "spike-and-slab" prior distribution, half of its probability is devoted to the null value $\theta_0$. If $\theta$ is equal to $\theta_0$, we use 1 bit to code it. The other half of the probability is uniformly distributed over a grid range which has the order of $\sqrt{n}$ positions. To encode that $\theta$ is not equal to $\theta_0$, we need $1 + \frac{1}{2}\log(n)$ bits where the

1 bit is to say that the estimator is not $\theta_0$ and the $\frac{1}{2}\log(n)$ bits is to specify the grid location. Since the term $\frac{1}{2}\log(n)$ dominates the main cost when coefficient is significant different from the null value, the coding scheme under "spike-and-slab" prior distribution is equivalent a BIC penalty when a feature is added to the model. This prior distribution means that, if the coefficient is not $\theta_0$, it will be any place in the grid range.

In the universal prior, like the "spike-and-slab" prior distribution, half of its probability is also devoted to the null value $\theta_0$. But instead of assuming a uniform distribution, the universal prior concentrates more of the other half probability near $\theta_0$ and let it decays slowly. Therefore, if $\theta$ is not equal to $\theta_0$, the coding scheme under the universal prior assumption uses $2 + \log^+ |\tilde{z}| + \log^+ \log^+ |\tilde{z}|$ bits to code it. This distribution assumption makes more sense in hard problems of feature selection where beneficial features are marginal significant. Because $\tilde{z}$ is relative small in such hard problems, the 2 bits will dominate the coding cost in such hard problem. This leads to a coding scheme which resembles AIC. In this book, we assume the "spike-and-slab" prior distribution and use 2 bits to code a coefficient.

## 4.2.3   Comparison of Three Coding Schemes

We compare the reduction of total description length, $\Delta S_{jM}^k$, in three coding schemes: a partially-dependent MIC as described above, a fully-dependent MIC, and independent regressions. The three coding schemes have different assumptions about the beneficial feature distribution in data. We define *beneficial* or *spurious* features as those which, if added to the current model, would or would not reduce prediction error, respectively, on a hypothetical infinite large test data set [47].

Previous sections of this book described the partially-dependent MIC coding scheme, which assumes regressions share some common feature information or structure. In this setting, feature evaluation considers the information that the whole response matrix can provide; if a feature is important, it will be added into a number $k$ ($1 \leq k \leq h$) of regressions.

Like partially-dependent MIC, fully-dependent MIC considers the whole response matrix when evaluating features. However, it assumes that all regressions share common features. So, if a feature is important, fully-dependent MIC will add it into all regressions, i.e., $k = h$. In coding, fully-dependent MIC differs from partially-dependent MIC is that it does not need to spend bits to specify how many and which regressions will add a feature, i.e., $\ell_H$ bits in equation 4.8 are saved. However, fully-dependent MIC may have to code more coefficients than partially-dependent MIC.

In the independent regressions coding scheme, each regression selects feature independently; the model will cost $\log(m)$ bits to code the feature index and $\ell_\Theta$ bits to code the coefficient. If summation of the two cost in bits is less than the bits saved by the increase of data likelihood given the new model with the feature, and the difference is beyond a specified threshold, this feature will be added to model. This coding scheme assumes that true regression models do not share significant common feature information or structure.

For simplicity,[3] we compare the three coding schemes in stepwise feature selection. For example, consider the case where we evaluate a feature, $x_j$, which is a beneficial feature for $k$ regressions and a spurious feature for the other $h - k$ regressions; for fully-dependent MIC, the $h - k$ regressions which add the spurious feature save negligible bits in encoding the residual error; partially-dependent MIC and independent regressions are assumed not to add any spurious feature; if the feature is added, the three methods save the same number of bits in encoding residual errors, $\Delta S_E^k$; for MIC coding schemes, $\ell_I = \log(m)$ and $\ell_\Theta = 2$; $m \gg h \gg 1$.

The results of comparison in theory are presented in Table 4.2. Independent regressions and partially-dependent MIC are the best and second best coding schemes

---

[3]The comparisons in other cases are more complex, but have similar results.

| when $k = 1$ | | |
|---|---|---|
| how to compare | (approximate) value of difference | $< 0, > 0,$ or $= 0$ |
| $\Delta S_{indep}^k - \Delta S_{partial.mic}^k$ | $\log(h)$ | $> 0$ |
| $\Delta S_{partial.mic}^k - \Delta S_{full.mic}^k$ | $2h - \log(h)$ | $> 0$ |
| $\Delta S_{indep}^k - \Delta S_{full.mic}^k$ | $2h$ | $> 0$ |
| when $k = \frac{h}{2}$ | | |
| how to compare | (approximate) value of difference | $< 0, > 0,$ or $= 0$ |
| $\Delta S_{partial.mic}^k - \Delta S_{indep}^k$ | $\frac{h-2}{2} \log(m) - \log\left(\frac{h^2 \times \binom{h}{h/2}}{4}\right)$ | $> 0$ |
| $\Delta S_{partial.mic}^k - \Delta S_{full.mic}^k$ | $h - \log\left(\frac{h^2 \times \binom{h}{h/2}}{4}\right)$ | $\approx 0$ |
| $\Delta S_{full.mic}^k - \Delta S_{indep}^k$ | $h \log(\sqrt{m})$ | $> 0$ |
| when $k = h$ | | |
| how to compare | (approximate) value of difference | $< 0, > 0,$ or $= 0$ |
| $\Delta S_{partial.mic}^k - \Delta S_{indep}^k$ | $h \log(m) - 2\log(h)$ | $> 0$ |
| $\Delta S_{full.mic}^k - \Delta S_{partial.mic}^k$ | $2\log(h)$ | $> 0$ |
| $\Delta S_{full.mic}^k - \Delta S_{indep}^k$ | $h \log(m)$ | $> 0$ |

Table 4.2: **Comparison of three coding schemes.** $m \gg h \gg 1$; for MIC coding schemes, $\ell_I = \log(m)$ and $\ell_\Theta = 2$. Please see Section 4.2.3 for other assumptions. $\Delta S_{indep}^k$, $\Delta S_{partial.mic}^k$, and $\Delta S_{full.mic}^k$ are description length decreased (if the feature is added into $k$ regressions) by independent regressions, partially-dependent MIC, and fully-dependent MIC respectively. For clarity, the subscript $j$ is omitted.

when $k = 1$, and their difference is on the order of $\log(h)$. Fully-dependent MIC and partially-dependent MIC are the best and second best coding schemes when $k = h$, and their difference is also on the order of $\log(h)$. Partially-dependent MIC and fully-dependent MIC are the best coding schemes when $k = \frac{h}{2}$.

## 4.2.4 MIC Coding Scheme in Streamwise Feature Selection

Streamwise feature selection [47] is designed for settings in which the feature set size is unknown. In streamwise feature selection, features are considered sequentially for addition to a model, and either added to the model or discarded; streamwise feature selection only evaluates each feature once when it is generated. To exploit streamwise feature selection, we proposed a penalty-based streamwise regression algorithm, information investing, that, unlike fixed penalty methods such as AIC, BIC, and RIC, uses flexible penalties to produce simple and accurate models when $m$ is big.

In streamwise feature selection, *Bid Selection* determines how many bits, $bid_j$, one is willing to spend to code the fact of not adding a feature $x_j$.[4] Bid selection can be done optimally by keeping track of the bits available to cover future overfitting (that is, the *wealth* $w_j$). Asymptotically, $bid_j$ is also the probability of adding this feature. Therefore, in the multiple responses setting, streamwise regression has

$$\ell_I = -\log(bid_j), \tag{4.11}$$

and other parts of coding remain the same as for stepwise regression. In our experiments, streamwise information-investing was investigated in the multiple response setting.

## 4.3 Experimental Evaluation

The different methods were evaluated on a synthetic data set containing 100 observations, each with 1,000 features. The features are generated independently from a normal distribution, $N(0,1)$. There are 20 responses; each response $y_i$ has $m^* = 4$ beneficial features in its true model which is the sum of the 4 beneficial features:

$$f_i(x) = \sum_{j^*=1}^{m^*} a_{ij^*} x_{j^*}, \tag{4.12}$$

and the $i$th response is

$$y_i = f_i(x) + \epsilon_i \tag{4.13}$$

where $a_{ij^*}$ are generated from the normal distribution $N(0,1)$, and residual vector $\epsilon_i$ are generated from a normal distribution $N(0, 0.1)$.

---

[4]In order to avoid confusion with error terms, we use *bid*, instead of $\epsilon$ which was used in [47].

We generate features in three scenarios. In scenario one, all responses share two beneficial features, and the other two beneficial features are different between responses; in scenario two, all responses share all four beneficial features; in scenario three, each response has a different set of four beneficial features, without any beneficial feature same with other response's. Note that, besides above specifications, beneficial features are randomly distributed in each run.

Results of scenario one are presented in Table 4.3. For streamwise regressions, fully-dependent MIC and partially-dependent MIC have better results than independent regressions. Fully-dependent MIC adds more than 20 spurious features, but at least finds more beneficial features than independent regressions. For stepwise regressions, partially-dependent MIC and independent regressions have the same good results, and fully-dependent MIC have worse results by including more than 70 spurious features.

Results of scenario two (Table 4.4) show that if responses have strong dependencies by sharing all beneficial features, fully-dependent MIC and partially-dependent MIC give best results for both streamwise and stepwise regressions. In contrast, results of scenario three (Table 4.5) show that if responses don't share any beneficial feature, independent regressions have better results than fully-dependent MIC and partially-dependent MIC.

As the results shows, if there are features shared by different regressions, partially-dependent MIC may borrow strength between regressions, and select more beneficial features and produce more accurate models than doing regressions individually. The results are consistent with our theoretical analysis in Section 4.2.3. Fully-dependent MIC shows its superior performance only if responses share all beneficial features.

### 4.3.1   Comparison with alternate methods

One can also use PLS, neural networks, or Curds and Whey, none of which do feature selection. Each has disadvantages, PLS is a linear method, and without feature

71

| stream. | independent | | fully. MIC | | partially. MIC | |
|---|---|---|---|---|---|---|
| penalty | RIC | info. | RIC | info. | RIC | info. |
| features | 2.8 | 2.6 | 29.1 | 30.4 | 6.0 | 6.3 |
| false pos. | 0.2 | 0.3 | 25.8 | 27.0 | 2.5 | 2.7 |
| error | 0.63±0.06 | 0.70±0.08 | 0.58±0.05 | 0.56±0.05 | 0.43±0.03 | 0.42±0.02 |

| step. | independent | fully. MIC | partially. MIC |
|---|---|---|---|
| penalty | RIC | RIC | RIC |
| features | 4.1 | 79 | 4.5 |
| false pos. | 0.3 | 75.2 | 0.7 |
| error | 0.33±0.00 | 0.73±0.01 | 0.33±0.00 |

Table 4.3: **Streamwise and stepwise regressions on synthetic data with multiple responses. Scenario one: two beneficial features shared.** The number of feature selected, false positives, and out-of-sample error, averaged over 10 runs. "RIC" penalty means the coding scheme $\ell_I = -\log(m)$ is used; "info." penalty means the coding scheme $\ell_I = -\log(bid_j)$ is used in the information-investing algorithm. $n = 100$ observations, $m = 1,000$ features, $h = 20$ responses in data. Each response has $m^* = 4$ beneficial features in its true model, and the beneficial features are randomly distributed in the feature set; all responses' true models share 2 common beneficial features, and the other 2 beneficial features are different between models. $x \sim N(0, 1)$, $y_i$ is linear in $x$ with noise $\sigma^2 = 0.1$. Average null error = 2.14.

| stream. | independent | | fully. MIC | | partially. MIC | |
|---|---|---|---|---|---|---|
| penalty | RIC | info. | RIC | info. | RIC | info. |
| features | 2.6 | 2.6 | 4.0 | 4.0 | 4.0 | 4.0 |
| false pos. | 0.3 | 0.3 | 0.0 | 0.0 | 0.0 | 0.0 |
| error | 0.48±0.03 | 0.51±0.03 | 0.32±0.00 | 0.32±0.00 | 0.32±0.00 | 0.32±0.00 |

| step. | independent | fully. MIC | partially. MIC |
|---|---|---|---|
| penalty | RIC | RIC | RIC |
| features | 3.8 | 4.1 | 4.0 |
| false pos. | 0.4 | 0.1 | 0.0 |
| error | 0.34±0.00 | 0.32±0.00 | 0.32±0.00 |

Table 4.4: **Streamwise and stepwise regressions on synthetic data with multiple responses. Scenario two: four beneficial features shared.** The number of feature selected, false positives, and out-of-sample error, averaged over 10 runs. All responses share $m^* = 4$ common beneficial features in their true models, and the beneficial features are randomly distributed in the feature set. Average null error = 1.70. Otherwise the same model and notations as Table 4.3.

| stream. | independent | | fully. MIC | | partially. MIC | |
|---|---|---|---|---|---|---|
| penalty | RIC | info. | RIC | info. | RIC | info. |
| features | 3.2 | 3.0 | 32.1 | 38.0 | 2.9 | 3.2 |
| false pos. | 0.3 | 0.4 | 30.6 | 36.3 | 0.8 | 0.9 |
| error | 0.49±0.04 | 0.55±0.05 | 1.14±0.11 | 1.14±0.11 | 0.70±0.06 | 0.66±0.05 |

| step. | independent | fully. MIC | partially. MIC |
|---|---|---|---|
| penalty | RIC | RIC | RIC |
| features | 4.1 | 79 | 5.0 |
| false pos. | 0.3 | 76.5 | 1.5 |
| error | 0.33±0.00 | 1.39±0.13 | 0.35±0.00 |

Table 4.5: **Streamwise and stepwise regressions on synthetic data with multiple responses. Scenario three: no beneficial features shared.** The number of feature selected, false positives, and out-of-sample error, averaged over 10 runs. Each response has $m^* = 4$ beneficial features in its true model, and the beneficial features are randomly distributed in the feature set; no features are shared by true models, i.e., all of $h \times m^* = 20 \times 4 = 80$ beneficial features (for different models) are different with each other. Average null error = 1.75. Otherwise the same model and notations as Table 4.3.

selection it is hard to include interaction terms. Neural networks suffer the opposite problem, being nonlinear methods which, as such, may require more observations to fit data with many featurees than are available.

PLS regression [45, 20] reduces the dimension of feature matrix in simultaneous or multivariate regressions by constructing new features or latent variables which are linear combinations of raw features and explain most of the covariance between feature matrix and response matrix. MIC also can find linear combinations of raw features which are shared by different models by using principal components (from PCA) as features.

In the Curds and Whey procedure [7], separate least squares regressions are trained on all features for respective responses, and their prediction models are linearly combined to take advantage of correlations between prediction tasks and obtain a more accurate prediction model for each response. In this procedure, information sharing between different tasks occurs after different models of prediction tasks have been built. In contrast, MIC shares information when select features and build

|          | scenario 1 | scenario 2 | scenario 3 |
|----------|-----------|-----------|-----------|
| null error | 2.14±0.16 | 1.70±0.15 | 1.75±0.13 |
| PLS error | 2.04±0.15 | 1.62±0.14 | 1.68±0.12 |

Table 4.6: **PLS results on the same synthetic data.** Same synthetic data and scenarios used in Table 4.3, 4.4, and 4.5.

models, and its motivation is consistent with the central idea of multi-task learning: sharing what is learned by different tasks while tasks are trained in parallel. Multi-task learning [8] differs from Curds and Whey by sharing internal structure learned by the models. It improves generalization performance of target task by adding extra tasks as extra outputs of an artificial neural net. However, like Curds and Whey, multi-task learning does not do feature selection.

Unlike PLS, neural networks, and Curds and Whey, MIC includes feature selection. In addition, MIC makes it easier to explore (with heuristic methods as described in [47]) the space of pairwise interaction terms or other nonlinear combinations of features with higher order even if $m$ is big. As above comparison indicate, MIC's strength lies in selecting features by sharing information between different models in the setting of multiple response with a vast, or even infinite, set of potentially predictive features.

We evaluated PLS on the same three synthetic data sets used above. The results are given in Table 4.6. As the results show, the performances of PLS are much worse than those of MIC in corresponding scenarios.

## 4.4   Discussion

In this book, we generalized RIC to the case where a set of features are simultaneously considered for addition to a set of regressions.

One might think that using PCA to make the resposes orthogonal would eliminate the benefit of sharing strength between the features. This is not the case. Perhaps non-intutively, it can easily be the case that the $y$'s are uncorrelated, but strength can still be borrowed between them. For example, consider a model

$$y_1 = x_1 + x_2$$

$$y_2 = x_1 - x_2.$$

Here $y_1$ and $y_2$ are uncorrelated, yet when doing feature selection over $x_1$, $x_2$ and a multiple noise features (e.g., features which are not predictive of the response), one will do a better job by "borrowing strength" on the presence of the same features in each model. Note that there is no benefit to borrowing strength on their coefficients; one would gain power on estimating the coefficient of $x_1$, and loose and equal amount of power when estimating the coefficient of $x_2$.

# Chapter 5

# Conclusions and Future Work

## 5.1 Conclusions

As the book presents, streamwise regression has performance comparable to, and even superior to, stepwise regression when one has a vast set of potentially predictive features of which only a small number are expected to be useful in predictive models. When using one of the two adaptive complexity penalty methods, $\alpha$-investing and information-investing, which dynamically adjust the threshold on the error reduction required for adding a new feature, streamwise feature selection avoids over- or under-fitting even in the limit of infinite numbers of non-predictive features, whereas streamwise AIC and BIC over-fit and streamwise RIC under-fits. Adding additional features often hurts streamwise AIC, BIC, and RIC. In contrast, the additional features have little effect on $\alpha$-investing or information-investing. False discovery rate (FDR)-style bounds for $\alpha$-investing and information-investing were proved using a supermartingale.

Emperical evaluation results show that although streamwise feature selection is effected by feature order. Shuffling the data usually has a fairly small effect on performance, as long as one is not using heuristic search over, e.g. interaction terms.

When compared with standard machine learning methods such as SVMs, neural nets, and decision trees on the UCI data sets, most of which have moderate numbers of candidate features, streamwise regression has comparable or superior performance. When compared with the NIPS03 competition winners on the competition data sets which have large numbers of candidate features, streamwise regression shows comparable performance to the best methods that have been used.

Streamwise regression with information-investing is described and explained using the framework of Minimum Description Length (MDL). The two-part coding scheme of MDL gives us the intuition that, if a feature selection method has a more efficient coding scheme, especially for encoding the model(s), this method would be more effective in avoiding under- or over-fit and have better prediction performance. It is also the motivation of MIC method in the setting of multiple responses.

The MIC method shares information between different prediction tasks, thereby using fewer bits to code models, and producding more accurate predictions than building model for each task individually. Emperical valuation confirms that if there are beneficial features shared by different regression models, MIC methods (partially- and fully-dependent MIC) give performance superior to those of independent regressions.

## 5.2   Future Work

### 5.2.1   Auction-based Modeling using Multiple Experts

In feature selection problems, there are many possible functions for generating features. We call these feature generators "experts", and will propose an auction protocol for automatic construction of statistical models based on the features proposed by these experts. Our auction method will also have sound underpinnings in information theory and statistics, including provable bounds against over-fitting.

In our auction approach, *experts* propose features and bid to have them incorporated in a model. The *auctioneer* then tests the winning bid for inclusion in the model, and accepts the proposed feature if it is statistically significant (i.e., if it reduces the total description length of the model). Experts start with an initial endowment of wealth (bits that they can use to describe features). An expert's wealth is increased if it proposes a feature that is used, and loses wealth if a feature it proposes is tested and found not worth adding to the model. The auction thus strengthens good experts, while preventing over-fitting.

Experts (feature generators) can take many forms. Raw, untransformed, features can be manually grouped into classes, such as different kinds of or locations of sensors in a robot or chemical plant; different experts then propose measurements from different sets of sensors. Other *transformational* features can be generated by experts transforming the features either individually (e.g., taking the log or square root) or in groups, applying spatial or temporal filters, computing principle components, or computing products or ratios of other features. Note that these transformational experts can work on raw features or on other transformed features. They can also take advantage of which features have already been added to the model, for example only looking for transformations of features which have already been added to the model. Using knowledge about what features are already in the model is particularly important for, e.g., interaction terms, where one may not want to test the product of all pairs of raw features, but only to examine products in which at least one of the features is already in the model.

Auctions combine the advantages of both manual and automatic model construction. Domain expertise is used in the experts that generate features, while the advantages of automatic model construction are used in the search over large classes of potential models. Some of the expert features proposed will prove useful, others not; the model constructed takes advantage its ability to do feature selection. Feature selection, as opposed to, for example, a neural network model, produces much simpler and more understandable models.

78

## 5.2.2 Other Future Work

In order to assure that the streamwise feature selection package implemented in R is flexible and robust enough that many people can use it, we will tackle below problems.

We will use Bennett's inequality to determine the p-values which are used in *alpha*-investing. If a feature is sparse, t statistic can inflate the significant of the feature, that is, the p-value of the t statistic is extremely small. Bennett's inequality conservatively bounds the p-value.

We will handle with those features with missing values. For a categorical feature, we can treat the missing values as another category. For example, if a binary feature has missing values, we can replace these missing values with zeros and add a missing value indicator/feature. For a continuous feature, we can fill the missing values with the mean of the observed values and add a missing value indicator/feature.

# Bibliography

[1] Felix Abramovich, Y. Benjamini, D. Donoho, and Ian Johnstone. Adapting to unknown sparsity by controlling the false discovery rate. Technical Report 2000–19, Dept. of Statistics, Stanford University, Stanford, CA, 2000.

[2] H. Akaike. Information theory and an extension of the maximum likelihood principle. In B. N. Petrov and F. Csàki, editors, *2nd International Symposium on Information Theory*, pages 261–281, Budapest, 1973. Akad. Kiàdo.

[3] Rie H. Ando and Tong Zhang. A framework for learning predictive structures from multiple tasks and unlabeled data. *Journal of Machine Learning Research*, 6:1817–1853, 2005.

[4] Y. Benjamini and Y. Hochberg. Controlling the false discovery rate: a practical and powerful approach to multiple testing. *Journal of the Royal Statistical Society*, Series B(57):289–300, 1995.

[5] Peter Bickel and Kjell Doksum. *Mathematical Statistics*. Prentice Hall, 2001.

[6] Avrim Blum and Pat Langley. Selection of relevant features and examples in machine learning. *Artificial Intelligence*, 97(1-2):245–271, 1997.

[7] Leo Breiman and Jerome H. Friedman. Predicting multivariate responses in multiple linear regression. *Journal of the Royal Statistical Society. Series B*, 59(1), 1997.

[8] Rich Caruana. Multitask learning. *Machine Learning*, 28(1):41–75, 1997.

[9] D. L. Donoho and I. M. Johnstone. Ideal spatial adaptation by wavelet shrinkage. *Biometrika*, 81:425–455, 1994.

[10] S. Dzeroski and N. Lavrac. *Relational Data Mining*. Springer-Verlag, 2001.

[11] S. Dzeroski, L. D. Raedt, and S. Wrobel. Multi-relational data mining workshop. In *KDD-2003*, 2003.

[12] Peter Elias. Universal codeword sets and representations of the integers. *IEEE Trans. on Info. Theory*, 21:194–203, 1975.

[13] Francois Fleuret. Fast binary feature selection with conditional mutual information. *Journal of Machine Learning Research*, 5:1531–1555, 2004.

[14] D. P. Foster and E. I. George. The risk inflation criterion for multiple regression. *Annals of Statistics*, 22:1947–1975, 1994.

[15] D. P. Foster and R. A. Stine. Local asymptotic coding. *IEEE Trans. on Info. Theory*, 45:1289–1293, 1999.

[16] D. P. Foster and R. A. Stine. Adaptive variable selection competes with Bayes experts. Submitted for publication, 2004.

[17] D. P. Foster and R. A. Stine. Variable selection in data mining: Building a predictive model for bankruptcy. *Journal of the American Statistical Association (JASA)*, 2004. 303-313.

[18] D. P. Foster and R. A. Stine. Multiple hypothesis testing using the execess discovery count and alpha-investing rules. Technical report, Statistics Department, University of Pennsylvania, 2005.

[19] Ildiko E. Frank and Jerome H. Friedman. A statistical view of some chemometrics regression tools. *Technometrics*, 35(2):109–135, 1993.

[20] Paul H. Garthwaite. An interpretation of partial least squares. *Journal of the American Statistical Association*, 89(425):122–127, 1994.

[21] E. I. George. The variable selection problem. *Journal of the Amer. Statist. Assoc.*, 95:1304–1308, 2000.

[22] E. I. George and D. P. Foster. Calibration and empirical bayes variable selection. *Biometrika*, 87:731–747, 2000.

[23] R. Gilad-Bachrach, A. Navot, and N. Tishby. Margin based feature selection - theory and algorithms. In *Proc. 21'st ICML*, 2004.

[24] Isabelle Guyon. Nips 2003 workshop on feature extraction and feature selection. 2003.

[25] Isabelle Guyon, Steve Gunn, Masoud Nikravesh, and Lofti Zadeh. *Feature Extraction, Foundations and Applications*. Springer, 2006.

[26] Jean Jacod and Albert Shiryaev. *Limit Theorems for Stochastic Processes*. Springer-Verlag, NY, 2002.

[27] D. Jensen and L. Getoor. *IJCAI Workshop on Learning Statistical Models from Relational Data*. 2003.

[28] JMLR. Special issue on variable selection. In *Journal of Machine Learning Research (JMLR)*, 2003.

[29] I. M. Johnstone and B. W. Silverman. Needles and straw in haystacks: Empirical bayes estimates of possibly sparse sequences. *Annals of Statistics*, 32:1594–1649, 2004.

[30] Ron Kohavi and George H. John. Wrappers for feature subset selection. *Artificial Intelligence*, 97(1-2):273–324, 1997.

[31] Richard J. Larsen and Morris L. Marx. *An Introduction to Mathematical Statistics and Its Applications*. Prentice Hall, 2001.

[32] Jinyan Li and Huiqing Liu. Bio-medical data analysis. 2002.

[33] Radford M. Neal. *Bayesian Learning for Neural Networks*. Number 118 in Lecture Notes in Statistics. Springer-Verlag, 1996.

[34] Radford M. Neal. Defining priors for distributions using dirichlet diffusion trees. Technical Report 0104, Dept. of Statistics, University of Toronto, 2001.

[35] Radford M. Neal and Jianguo Zhang. Classification for high dimensional problems using bayesian neural networks and dirichlet diffusion trees. In *NIPS 2003 workshop on feature extraction and feature selection*, 2003.

[36] NIPS'03. Challenge results. 2003.

[37] A. Popescul and L. H. Ungar. Structural logistic regression for link analysis. In *KDD Workshop on Multi-Relational Data Mining*, 2003.

[38] A. Popescul and L. H. Ungar. Cluster-based concept invention for statistical relational learning. In *Proc. Conference Knowledge Discovery and Data Mining (KDD-2004)*, 2004.

[39] Jorma Rissanen. A universal prior for integers and estimation by minimum description length. *Annals of Statistics*, 11:416–431, 1983.

[40] Jorma Rissanen. Hypothesis selection and testing by the mdl principle. *The Computer Journal*, 42:260–269, 1999.

[41] Gideon Schwartz. Estimating the dimension of a model. *The Annals of Statistics*, 6(2):461–464, 1978.

[42] R. A. Stine. Model selection using information theory and the mdl principle. *Sociological Methods Research*, 33:230–260, 2004.

[43] John D. Storey. A direct approach to false discovery rates. *J. of the Royal Statist. Soc., Ser. B*, 64:479–498, 2002.

[44] Lyle H. Ungar, Jing Zhou, Dean. P. Foster, and Robert. A. Stine. Streaming feature selection using iic. In *AI&STAT'05*, 2005.

[45] Svante Wold, Michael Sjostrom, and Lennart Eriksson. Pls-regression: a basic tool of chemometrics. *Chemometrics and Intelligent Laboratory Systems*, 58(2):109–130, 2001.

[46] Jing Zhou, Dean P. Foster, Robert A. Stine, and Lyle H. Ungar. Streaming feature selection using alpha-investing. In *ACM SIGKDD'05*, 2005.

[47] Jing Zhou, Dean P. Foster, Robert A. Stine, and Lyle H. Ungar. Streamwise feature selection. *Journal of Machine Learning Research*, 7:1861–1885, 2006.

www.ingramcontent.com/pod-product-compliance
Lightning Source LLC
LaVergne TN
LVHW080102070326
832902LV00014B/2367